本书由中国科学院数学与系统科学研究院资助出版

数学 24/7

厨房中的数学

〔美〕海伦·汤普森 著

李静科 王丽平 译

科 学 出 版 社

北 京

图字：01-2015-5627号

内 容 简 介

厨房中的数学是"数学生活"系列之一，内容涉及食物的度量、食材的购买、食谱的制作、营养成分分析及能量的计算，同时介绍不同度量单位之间的换算及比例分配，让青少年在学校学到的数学知识应用到与食物制作有关的多个方面中。

本书适合作为中小学生的课外辅导书，也可作为中小学生的兴趣读物。

Copyright © 2014 by Mason Crest, an imprint of National Highlights, Inc. All rights reserved. No part of this publication may be reproduced or transmitted in any form or by any means, electronic or mechanical, including photocopying, recording, taping or any information storage and retrieval system, without permission from the publisher.
The simplified Chinese translation rights arranged through Rightol Media.
（本书中文简体版权经由锐拓传媒取得Email:copyright@rightol.com）

图书在版编目（CIP）数据

厨房中的数学/（美）海伦·汤普森（Helen Thompson）著；李静科，王丽平译.—北京:科学出版社,2018.5
（数学生活）
书名原文：Culinary Math
ISBN 978-7-03-056358-3

Ⅰ.①厨… Ⅱ.①海… ②李… ③王… Ⅲ.①数学-青少年读物 Ⅳ.①O1-49

中国版本图书馆CIP数据核字（2018）第011938号

责任编辑:胡庆家 / 责任校对:邹慧卿
责任印制:肖 兴 / 封面设计:陈 敬

科学出版社 出版
北京东黄城根北街16号
邮政编码：100717
http://www.sciencep.com

北京汇瑞嘉合文化发展有限公司 印刷
科学出版社发行 各地新华书店经销

*

2018年5月第 一 版　开本：889×1194　1/16
2018年5月第一次印刷　印张：4 1/4
字数：70 000

定价：98.00元(含2册)
（如有印装质量问题，我社负责调换）

引　　言

你会如何定义数学？它也许不是你想象的那样简单。我们都知道数学和数字有关。我们常常认为它是科学，尤其是自然科学、工程和医药学的一部分，甚至是基础部分。谈及数学，大多数人会想到方程和黑板、公式和课本。

但其实数学远不止这些。例如，在公元前5世纪，古希腊雕刻家波留克列特斯曾经用数学雕刻出了"完美"的人体像。又例如，还记得列昂纳多·达·芬奇吗？他曾使用有着赏心悦目的尺寸的几何矩形——他称之为"黄金矩形"，创作出了著名的画作——蒙娜丽莎。

数学和艺术？是的！数学对包括医药和美术在内的诸多学科都至关重要。计数、计算、测量、对图形和物理运动的研究，这些都被融入到音乐与游戏、科学与建筑之中。事实上，作为一种描述我们周围世界的方式，数学形成于日常生活的需要。数学给我们提供了一种去理解真实世界的方法——继而用切实可行的途径来控制世界。

例如，当两个人合作建造一样东西时，他们肯定需要一种语言来讨论将要使用的材料和要建造的对象。但如果他们建造的过程中没有用到一个标尺，也不用任何方式告诉对方尺寸，甚至他们不能互相交流，那他们建造出来的东西会是什么样的呢？

事实上，即便没有察觉到，但我们确实每天都在使用数学。当我们购物、运动、查看时间、外出旅行、出差办事，甚至烹饪时都用到了数学。无论有没有意识到，我们在数不清的日常活动中用着数学。数学几乎每时每刻都在发生。

很多人都觉得自己讨厌数学。在我们的想象中，数学就是枯燥乏味的老教授做着无穷无尽的计算。我们会认为数学和实际生活没有关系；离开了数学课堂，在真实世界里我们再不用考虑与数学有关的事情了。

然而事实却是数学使我们生活各方面变得更好。不懂得基本的数学应用的人会遇到很多问题。例如，美联储发现，那些破产的人的负债是他们所得收入的1.5倍左右——换句话说，假设他们年收入是24000美元，那么平均负债是36000美元。懂得基本的减法，会使他们提前意识到风险从而避免破产。

作为一个成年人，无论你的职业是什么，都会或多或少地依赖于你的数学计算能力。没有数学技巧，你就无法成为科学家、护士、工程师或者计算机专家，就无法得到商学院学位，就无法成为一名服务生、一位建造师或收银员。

体育运动也需要数学。从得分到战术，都需要你理解数学——所以无论你是想在电视上看一场足球比赛，还是想在赛场上成为一流的运动员，数学技巧都会给你带来更好的体验。

还有计算机的使用。从农庄到工厂、从餐馆到理发店，如今所有的商家都至少拥有一台电脑。千兆字节、数据、电子表格、程序设计，这些都要求你对数学有一定的理解能力。当然，电脑会提供很多自动运算的数学函数，但你还得知道如何使用这些函数，你得理解电脑运行结果的含义。

这类数学是一种技能，但我们总是在需要做快速计算时才会意识到自己需要这种技能。于是，有时我们会抓耳挠腮，不知道如何将学校里学的数学应用在实际生活中。这套丛书将助你一马当先，让你提前练习数学在各种生活情境里的运用。这套丛书将会带你入门——但如果想掌握更多，你必须专心上数学课，认真完成作业，除此之外再无捷径。

但是，付出的这些努力会在之后的生活里——几乎每时每刻（24/7）——让你受益匪浅！

目　　录

引言
1. 食品中的数学　　　　　　　　　　　　　　　　1
2. 在食品店估算　　　　　　　　　　　　　　　　3
3. 用优惠券买食品　　　　　　　　　　　　　　　5
4. 厨房中的量度　　　　　　　　　　　　　　　　7
5. 厨房中的分数　　　　　　　　　　　　　　　　9
6. 食谱中的数学　　　　　　　　　　　　　　　　11
7. 多人食谱中的数学　　　　　　　　　　　　　　13
8. 批量烹饪　　　　　　　　　　　　　　　　　　15
9. 厨房中时间的利用　　　　　　　　　　　　　　17
10. 液体中的数学　　　　　　　　　　　　　　　　19
11. 营养中的数学　　　　　　　　　　　　　　　　21
12. 大于和小于：食品标签上的数学　　　　　　　　23
13. 计算卡路里　　　　　　　　　　　　　　　　　25
14. 你需要多少卡路里？　　　　　　　　　　　　　27
15. 小结　　　　　　　　　　　　　　　　　　　　29
参考答案　　　　　　　　　　　　　　　　　　　　31

Contents

INTRODUCTION
1. GROCERY MATH   37
2. ESTIMATING AT THE GROCERY STORE   39
3. GROCERY SHOPPING WITH COUPONS   40
4. KITCHEN MEASUREMENTS   42
5. FRACTIONS IN THE KITCHEN   43
6. RECIPE MATH   44
7. MORE RECIPE MATH   45
8. COOKING IN BATCHES   47
9. USING TIME IN THE KITCHEN   48
10. LIQUID MATH   50
11. NUTRITION MATH   51
12. GREATER THAN AND LESS THAN:
    FOOD LABEL MATH   52
13. COUNTING CALORIES   54
14. HOW MANY CALORIES DO YOU NEED?   55
15. PUTTING IT ALL TOGETHER   56
ANSWERS   57

1
食品中的数学

拉马尔去了烹饪学校,他正在那里学习如何成为一名厨师。今天,他计划为他的5个朋友准备一顿特别的饭。因为他们将要庆祝一位朋友的生日,所以拉马尔想准备朋友们最喜欢的食物:奶酪通心粉、汉堡、沙拉和巧克力蛋糕。拉马尔已经选好了食谱,并根据他手头上现有的食材,做了一个食品清单。

根据他的食品清单,他还需要购买下面这些食材:

1磅通心粉
1磅奶酪
1夸脱牛奶
3磅碎牛肉
1棵生菜
3个西红柿
1罐番茄酱
1瓶芥末
1打(12个)汉堡面包
5磅面粉

5磅糖
1包烘焙巧克力
1打鸡蛋
1罐发酵粉
盐
1瓶沙拉酱
1瓶植物油
1磅黄油
细砂糖

拉马尔一直很节省，并且他只有55美元可以花在这顿饭上。当他到达食品店时，他发现食材的价格如下：

通心粉：每磅0.99美元
奶酪：每磅3.69美元
牛奶：每夸脱1.53美元
碎牛肉：每磅1.75美元
生菜：每棵1.15美元
西红柿：每磅2.50美元
番茄酱：每瓶1.87美元
芥末：每瓶0.99美元
汉堡包：3.75美元6个
面粉：每5磅3.10美元

糖：每5磅2.95美元
烘焙巧克力：每包4.75美元
鸡蛋：每打2.99美元
发酵粉：每罐3.19美元
盐：每箱2.10美元
沙拉酱：每瓶2.75美元
食用油：每瓶3.25美元
黄油：每磅2.80美元
细砂糖：每包2.15美元

他将花多少钱呢？

为了找到答案，你需要将每一项的费用加起来。不过，首先需要将每一项的份量都乘以单价。

如果拉马尔需要3磅的碎牛肉，那么需要将单价乘以3：

$$1.75 \times 3 =$$

如果拉马尔需要一打汉堡包，那么需要将6个汉堡包的价格乘以2：

$$3.75 \times 2 =$$

如果他需要3个西红柿，那么他需要知道它们多重。平均每个西红柿大约重4盎司，这意味着4个西红柿重约1磅(16盎司)。但是，拉马尔只需要3个西红柿。因此，我们需要找到下面式子的答案：

$$3/4 \times 2.50 =$$

现在，我们需要将每一项加起来。
拉马尔有足够的钱吗？
他还有剩余的钱吗？
如果够用的话，剩余多少钱？
如果不够用的话，他还需要多少钱？

2
在食品店估算

当拉马尔进入食品店时,他决定还要做一个水果沙拉。他买了一些香蕉和草莓。辣椒看起来也很好看,他想在油拌沙拉里放一些辣椒。

但他不确定是否有足够的钱。香蕉花了79美分,草莓花了3.89美元,辣椒花了33美分。他没有带计算器来帮他像刚才那样将所有量加起来。对他来说,用大脑来计算,数字太多了。

那么,当他到收银台时,拉马尔怎样才能避免钱不够的尴尬呢?他不想等到结账时才发现自己没有足够的钱来购买所需要的食品!

用心算来估计将要为这些食品支付多少钱是一个好办法。你可以通过四舍五入到最接近的金额。

填写下面的表。前两项已经为你做好了。当你完成后,就可以看出拉马尔是否还有钱买这些额外的食品。

项目	单价	你的估计/美元	累计/美元
2磅通心粉	0.99美元/磅	2.00	2.00
1磅奶酪	3.69美元/磅	4.00	6.00
1夸脱牛奶	1.53美元/夸脱		
3磅碎牛肉	1.75美元/磅		
3个西红柿(3/4磅)	2.50美元/磅		
1瓶番茄酱	1.87美元/瓶		
1罐芥末	0.99美元/罐		
1打汉堡包	3.75美元/6个		
5磅面粉	3.10美元/5磅		
5磅糖	2.95美元/5磅		
烘焙巧克力	4.75美元/包		
鸡蛋	2.99美元/打		
发酵粉	3.19美元/罐		
盐	2.10美元/箱		
沙拉酱	2.75美元/瓶		
食用油	3.25美元/瓶		
1磅黄油	2.80美元/磅		
细砂糖	2.15美元/包		
香蕉	0.79美元/把		
草莓	3.89美元/包		
辣椒	0.33美元/个		

拉马尔有足够的钱吗？

3
用优惠券买食品

拉马尔在购物时,遇到了他的朋友凯莉也在购物。凯莉与拉马尔分享了她的一些优惠券,这样他在结账时就可以少花一些钱了。

优惠券通常印在报纸和杂志上。只需要剪下优惠券,并带着它们去商店。有时候,你可以在网上找到优惠券,然后打印出来就可以了。结账时,把优惠券给收银员,她就会从相应的商品中减去一些钱。

如果拉马尔使用下一页的优惠券,将为他省多少钱呢?

百分比意味着占100的份额,所以25%意味着25/100。这就等同于四分之一。(你可以这样记,1美元有4个25美分,于是每四分之一相当于25美分。)所以如果某件商品降价25%,这就意味着首先需要搞清楚它的定价的四分之一是多少。你可以通过将黄油的定价除以4,然后从定价中减去这个数字。

所得到的等式如下:

$$2.80 美元 - (2.80 美元 \div 4) =$$

(请记住,要先解括号内的部分。)

50%也就是50/100,即一半。如果某个商品降价50%,要搞清楚它要花费多少钱,一个简单的方法是,搞清楚如果你从它的定价中减掉一半将是多少。在这种情况下,如果把盐的价格除以2,就会知道用优惠券拉马尔买盐要花多少钱。

20%也就是20/100,100是20的几倍?

一旦你知道,你就会知道用几去除面粉的价格。

然后再从价格中减去上面所得的商。

你的答案是什么?

另一种方式,你可以计算出百分比:使用小数。对此,请记住,将小数点移动两位,所以20%=0.20。你可以将面粉的价格乘以0.20(3.10 × 0.20),会发现20%相当于多少。然后从3.10美元中减去这个数便会知道减价20%后是多少。

无论选择哪种方式,你都会得到相同的答案。不妨选择计算更容易的方式。

如果拉马尔3张优惠券都用上,他共节省多少钱?

4
厨房中的量度

现在拉马尔已经买了他需要的所有食材，准备开始做饭。他将这些食材放在灶台上，并根据他的食谱做准备。但是他的食谱要求的度量有些在食品包装上并未标明。他需要从一个度量单位转换成另一种，以确保他用适量的配料。

如果使用下面的图表，应该能够相当容易地从一种度量单位转换成另一种。为了找到正确的答案，你将需要用乘法或除法。

1杯磨碎的奶酪=8盎司

3茶匙=1汤匙   16汤匙=1杯

1杯=8盎司=1/2品脱   2杯=16盎司=1品脱

4杯=32盎司=2品脱=1夸脱   2夸脱=64盎司=1/2加仑

4夸脱=128盎司=1加仑   1方块烘焙巧克力=1盎司

1条黄油=8汤匙=1/2杯=4盎司

1. 拉马尔的奶酪通心粉食谱需要8盎司磨碎的奶酪。他需要多少杯（奶酪）？

2. 如果食谱中需要1 1/2杯牛奶，合多少盎司？

3. 食谱要求1/4杯黄油。这将是1条黄油的多少？又将是多少汤匙呢？

4. 拉马尔的巧克力蛋糕食谱需要16盎司面粉，合多少杯？

5. 蛋糕需要4盎司的烘焙巧克力，合多少方块？

6. 如果拉马尔需要3汤匙泡打粉，将是多少茶匙？

7. 2夸脱水将是几杯？

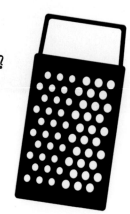

5
厨房中的分数

食谱通常告诉你按照食谱可以做出多少人的份量。但有时候你需要做出不同于食谱上的份量。当发生这种情况时，必须调整配方中各种材料的用量。尤其当你烤东西的时候，不能只是多加一种配料就期待做出美味的食物。蛋糕、饼干和面包的食谱就像化学公式：你需要保持材料的量完全正确，否则蛋糕、饼干或面包的发酵不能令人满意。它可能会变得扁平而硬——也可能发酵太过，充满气泡。

所以如果你想加倍食谱中的份量，就需要将每一种材料分量乘以2。如果你想要食谱份量的一半，就需要使用每一种材料分量的刚好一半。在下一页上，你会发现一些例子。

如果你想把需要 2 1/2 杯面粉的食谱份量翻倍,首先需要做出假分数:

2 = 2/1

2 1/2 = 5/2

然后把它们相乘:

2/1 × 5/2 = 10/2 = 5

所以现在你知道需要5杯面粉来加倍食谱份量。

这一次假设食谱要求为 3 1/2 杯面粉,但是你只想做食谱一半的份量。这时乘以1/2。当你把3 1/2变成假分数时,会得到7/2,所以

1/2 × 7/2 = 7/4 = 1 3/4

因此要做食谱一半的量,你需要使用 1 3/4 杯面粉。得到相同答案的一个简单方法是要记住这些规则:

双份食谱:要做多一倍份量,每种配料的用量乘以2。

减半的食谱:只做一半份量,每种配料的用量除以2。

但记住,假如你正在使用分数,你做乘法之前需要先把它们变成假分数!

现在看看你能否找出这些问题的答案:

1. 如果食谱可做4人份——但是你需要12人份?你需要用食谱中的数字做乘法还是除法?你会用什么数字?

2. 如果食谱可做15人份,但你只想要5人份。你需要用食谱中的数字做乘法还是除法?你会用什么数字?

6
食谱中的数学

拉马尔正在为他的朋友们做奶酪通心粉。他想要提供6人的食物并且足够每人一份——但他正在使用的食谱份量只是4人份。拉马尔需要做什么?

更加棘手的是,他需要做出比食谱上多1/2的量。因此,首先他需要将每种材料的用量除以2。然后他需要将该用量加到原始用量上。

在下一页上,更改食谱,使它变成6人份而不是4人份。这是你首先要做的事情。

4人份	+4的1/2份	=6人份
1包通心粉	+1/2包	=1 1/2包通心粉
2个鸡蛋		
2杯牛奶		
2汤匙融化的黄油		
2 1/2杯切碎的奶酪		

可是，转念一想，拉马尔决定他想要有足够的奶酪通心粉，使得每个人都可以有第二份。这意味着他反而将需要做12人份的。你可以填写以下图表，改变材料的量，以使他这一次可以做出12人份吗？

4人份	×3	=12人份
1包通心粉	×3	=3包通心粉
2个鸡蛋		
2杯牛奶		
2汤匙融化的黄油		
2 1/2杯切碎的奶酪		

7
多人食谱中的数学

拉马尔还需要烤巧克力蛋糕。他决定邀请更多的人来享用甜点,所以他要确定蛋糕大到足以供应20人。不过,他打算使用的食谱只是8人份的。他如何才能计算出一个足够这么多人享用的蛋糕食谱?

他有几个选项:可以用20除以8,结果为2余数为4。因为4是8的一半,同样结果就是2 1/2,

$$20 \div 8 = 2\ 1/2$$

现在,他需要将食谱中的每种材料的分量乘以2 1/2。

但拉马尔想让自己做得更容易些。他转念决定将食谱增至3倍。这反而意味着他要将食谱中的每种材料的份量乘以3。乘以一个整数是比较容易的。

现在他可以把蛋糕切成稍大一点的片,仍然刚好足够每个人。或者他可以为自己留下4块!

填写下页的表,计算出假如他将食谱增至3倍,食谱中每种材料需要的份量。

原蛋糕食谱	翻3倍
1杯面粉	
1杯糖	
3块方形的烘烤巧克力	
1茶匙泡打粉	
3/4茶匙苏打粉	
1/2汤匙盐	
1/2杯牛奶	
1/4杯植物油	
1个鸡蛋	
1茶匙香草精	
1/2杯开水	
1杯中筋面粉	

现在拉马尔将需要确保他有足够的糖霜覆盖蛋糕。他可能不需要精确的3倍之多，但蛋糕会更大，所以将需要更多的糖霜。他决定要将糖霜食谱配方翻倍。你能算出他所需要的数量吗？

原糖霜食谱	两倍糖霜食谱
1 1/2杯黄油	
1杯可可粉	
5杯细砂糖	
1/2杯牛奶	
2茶匙香草精	

拉马尔现在需要多少条黄油制造糖霜？假如需要，请参考第6节。

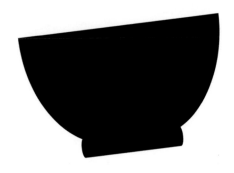

8
批量烹饪

拉马尔在桌子上放置了做汉堡包的小烧烤架。不过,烤架上一次只能烤4个汉堡包。如果拉马尔想要做足够的汉堡包使得每人2个,他需要12个汉堡包,因为预期的是6个人。这意味着他将需要一次烤一批。他需要烤几批?

$$12 \div 4 = 3$$

他需要烤3批。

如果他想要让每个人都有3个汉堡包，他将需要烤几批？

现在拉马尔意识到烤这么多汉堡包会花掉很长时间。他将需要留出足够时间来烤出所有批次的汉堡包。

如果烤一批需要8分钟，烤好3批要用多久？

如果他要烤18个汉堡包，让每个人都有3个汉堡包，他需要为此留出多长时间？

批次练习

请填写下面的图表作为进一步练习。

	几批？	烹饪时间
你想要烤8片吐司面包，但是烤面包机一次只能放2片面包。烤每一批需要3分钟		
你烤饼干需要烤12分钟。食谱是36个饼干，但烤盘内一次只能容纳12个饼干		

9
厨房中时间的利用

拉马尔的朋友乔治停下来并对他说,他想帮忙准备食物。拉马尔让乔治切做沙拉要用到的西红柿——但乔治切得太慢了,拉马尔担心西红柿无法切完。拉马尔做了个深呼吸并看了一下表,计算着他现在还剩多少时间。

现在是下午4:00,他的朋友们将于6:30到达。这意味着他就剩下两个半小时的准备时间了。

奶酪通心粉已经做好放入了烤箱。蛋糕也就只差烘焙了,但他不能将蛋糕和通心粉同时放入烤箱(此外,两者需要不同的温度)。

下表显示了拉马尔还需做的事情，并列出了每项事情所需的时间。

待做事项	所需时间
烤蛋糕	35分钟
冷却蛋糕	10分钟
做糖霜	10分钟
给蛋糕上糖霜	10分钟
烤奶酪通心粉	45分钟
做凉拌沙拉	10分钟
做水果色拉	10分钟
布置桌面	10分钟
烤汉堡包	24分钟

如果拉马尔以上任务都要做，一次做一件事情，他需要花多长时间完成？

他能在客人到来之前完成吗？

好消息是——在他们使用烤箱的同时，拉马尔和乔治还可以做其他的工作！

拉马尔决定最后烤汉堡包比较好，这样将尽可能地保持汉堡包的热度。他先烤蛋糕，这样在他给蛋糕上糖霜之前，蛋糕会有冷却的时间。

所以，如果拉马尔先烤蛋糕，在烤完蛋糕后放入奶酪通心粉，那奶酪通心粉需要烤多长时间才能从烤箱中取出？

在烤蛋糕和烤通心粉时，拉马尔和乔治是否有时间去做其他的事情呢？

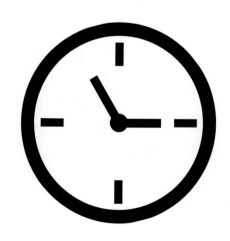

10
液体中的数学

拉马尔的朋友罗塞尔顺道来做客并帮助他和乔治一同准备食物。她带了一些食材：一袋燕麦饼干，因为凯莉不喜欢巧克力，还有几瓶苏打水和果汁。她和拉马尔决定混合苏打水和果汁做鸡尾酒。

罗塞尔买了5瓶10盎司姜汁啤酒，2加仑蔓越莓果汁和半加仑苹果汁。如果把这些混合在一起，她能得到多少杯鸡尾酒？

回顾第6节并填写这个表格：

	多少杯？
5瓶10盎司的苏打水=	
2加仑的蔓越莓果计=	
1/2加仑的苹果计=	
合计	

公制度量是什么？

公制度量是测量液体的另一种方法。
　　1加仑=3.8升
　　1夸脱=0.9升
　　1杯=0.2升

11
营养中的数学

拉马尔的朋友瑞提斯来了,他想弄清楚这顿饭里有多少营养成分。瑞提斯也打算去烹饪学校学习,但他比较关注营养方面的问题。他向朋友们解释说,每个人都需要在饮食中均衡不同营养成分的量。

罗塞尔想知道他们准备的这顿饭是否容易让人发胖,这些食物是否是健康食品。瑞提斯解释说,卡路里、脂肪、盐和糖都是决定一种食物是否健康的重要因素。他还说,不饱和脂肪酸比饱和脂肪酸更好。他拿出笔记本电脑并找到一个网页,在这个网页输入不同的食物便可查找出其所对应的营养值。

下列表格显示了他得到的结果。大部分数据对应的单位是克和毫克。1克约为一个回形针那么重。1毫克更轻，差不多等于一片雪花那么重。

1份	热量/卡路里	糖/克	脂肪/克	饱和脂肪/克	钠（盐）/毫克
奶酪通心粉	200	1	6	2	1027
汉堡	230	0	15	6	64
凉拌沙拉	25	0	0	0	0
水果沙拉	50	7	0	0	0
鸡尾酒	150	36	0	0	0
巧克力蛋糕	236	32	10	3	214

1. 如果拉马尔的朋友仅吃一份的话，他这一餐有多少卡路里？
2. 如果他们吃两份有多少卡路里？
3. 这一份餐中含多少糖（单位：克）？同下表的每日推荐所需量相比怎样？
4. 钠是盐的另一个名字。这一份餐中含有多少盐（单位：毫克）？
5. 当考虑到卡路里、脂肪、糖和盐时，你认为这餐是否健康呢？

每日饮食推荐

	男性	女性	儿童	9-13岁男孩	14-18岁男孩	9-13岁女孩	14-18岁女孩
脂肪	每日卡路里的20%~35%	每日卡路里的20%~35%	每日卡路里的25%~35%	每日卡路里的25%~35%	每日卡路里的25%~35%	每日卡路里的25%~35%	每日卡路里的25%~35%
饱和脂肪	少于每日卡路里的10%	少于每日卡路里的10%	少于每日卡路里的10%	少于每日卡路里的10%	少于每日卡路里的10%	少于每日卡路里的10%	少于每日卡路里的10%
钠	少于2300毫克	少于2300毫克	少于1900毫克	少于2200毫克	少于2300毫克	少于2200毫克	少于2300毫克
糖	37.5克	25克	12克	12克	12克	12克	12克

资料来源：美国食品药品管理局和美国心脏协会。

12
大于和小于：
食品标签上的数学

罗塞尔指出：在瑞提斯上网查找东西时，遗漏了色拉酱。在瑞提斯再次打开他的电脑之前，拉马尔告诉他的朋友，很多有标签的食物会注明其营养成分并指出是否属于健康食品。类似色拉酱调料类的食品常为瓶装，其他盒装或袋装食品的标签上也给出了很多营养成分的信息。

你将发现几乎所有带包装的食品上都有标签，标注在有一堆数字的黑白图表中。你每天需要100%的各种营养成分以达到最健康的状态。每日营养成分的百分比告诉你所摄取的营养成分有多接近使你要达到最健康状态所需的摄入量。

营养成分表	每份（30克）	每天营养素参考值%	每份（30克）	每天营养素参考值%
2茶匙份量（30克）	脂肪 6克	9%	碳水化合物 3克	1%
每瓶份量约12茶匙	饱和脂肪酸 3.5克	18%	食用纤维 0克	0%
热量 70	反式脂肪酸 0克		食糖 2克	
来自脂肪的热量 50	胆固醇 20毫克	7%	蛋白质 1克	
	钠 180毫克	7%		
	维生素A 6% ● 维生素C 2%		钙元素4% ● 铁元素0%	

*每日摄取值的百分比根据2000卡路里的饮食

将以上色拉酱的营养成分标签同罗塞尔买的下列饼干的营养成分标签相对比。请用<或>填写以下空格：

1. 哪种食物每份提供的脂肪较多？

   色拉酱_____饼干

2. 哪种食物每份所含卡路里更多？

   色拉酱_____饼干

3. 哪种食物含盐量更高？

   色拉酱_____饼干

4. 哪种食物每份含糖量更高？

   色拉酱_____饼干

你认为这些食物有利于健康吗？若是人们需要健康的饮食的话是否可以多吃这些食物？为什么呢？

营养成分表	每份（17克）	每天营养素参考值%	每份（17克）	每天营养素参考值%
每个饼干（17克）	脂肪 4.5克	7%	碳水化合物 7克	2%
每盒份量约12个饼干	饱和脂肪酸 2.5克	13%	食用纤维 1克	4%
热量80	反式脂肪酸 0克		食糖 4克	
来自脂肪的热量40	胆固醇 5毫克	2%	蛋白质 1克	
	钠 20毫克	1%		
	维生素A 0% ● 维生素C 0%		钙元素0% ● 铁元素0%	

*每日摄取值的百分比根据2000卡路里的饮食

13
计算卡路里

瑞提斯向他的朋友解释了不同营养成分及卡路里数值之间的关系。

瑞提斯说:"我们的身体需要从我们所吃的食物中摄取不同的营养成分。"我们需要均衡蛋白质、碳水化合物和脂肪的摄入量。蛋白质用于维持我们肌肉的健康,碳水化合物用来提供能量。我们同样需要有益的脂肪帮助我们吸收维生素并提供能量。"

蛋白质可以从肉类或奶制品中获得,比如奶酪和牛奶。碳水化合物可由面包、点心、水果和蔬菜中获得。脂肪可以从肉类、奶制品及很多零食中获取。

不同种类的营养成分具有不同的卡路里值。例如，每克蛋白质有4卡路里。如果你吃了含7克蛋白质的奶酪，你将从这些蛋白质中获取28卡路里。

$$7克蛋白质 \times 4卡路里/克 = 28卡路里$$

奶酪中其余的卡路里则来源于其他的营养成分。

碳水化合物每克也含4卡路里。例如，苏打水含糖量很高，糖是碳水化合物。一罐苏打水一般含有约40克的糖，从这些糖中总共可获取160卡路里。

$$40克糖 \times 4卡路里/克 = 160卡路里$$

苏打水中不含其他的营养成分，所以所有从中获得的卡路里都来源于糖。

每克脂肪所含的卡路里最多。每克脂肪有9卡路里。半个牛油果约有11克有益的不饱和脂肪，当你计算卡路里时，你会发现半个牛油果含有99卡路里。

$$11克脂肪 \times 9卡路里/克 = 99卡路里$$

回顾第11节并回答下列问题：

1. 一份奶酪通心粉中有多少卡路里来源于脂肪？
2. 一块巧克力蛋糕中有多少卡路里来源于糖？

现在回顾第12节的食品标签回答下列问题：

3. 一块饼干中有多少卡路里来自于蛋白质？
4. 一份色拉酱中有多少卡路里来源于脂肪？

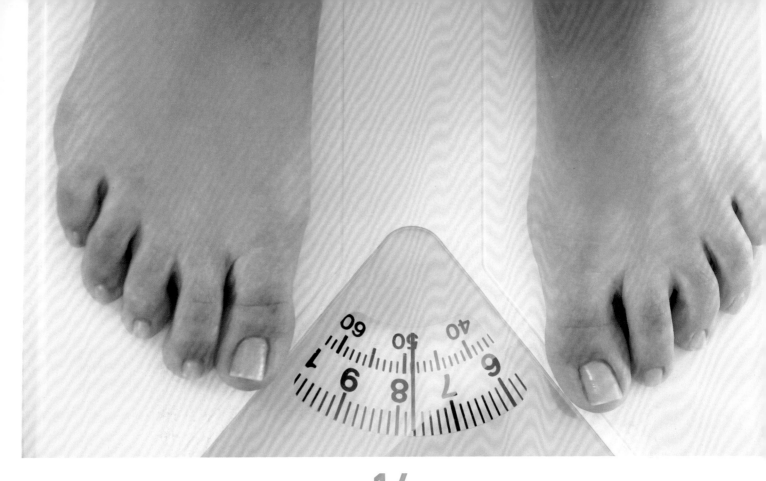

14
你需要多少卡路里?

拉马尔和他的朋友们吃了他准备的晚餐,他们在一起很开心。晚餐过后,凯莉和乔治打扫厨房,其他人则聚集在客厅。在把盘子放入洗碗机中清洗时,他俩又开始谈论卡路里了。乔治跟凯莉说,他可能超重了,想知道是否应该节食。凯莉向他解释说,你不能仅凭称上的数值来判断是否超重。每个人的体质是不同的。有些人骨骼较重,而有些人肌肉较多,一些人高,一些人矮,还有些人介于两者之间。

医生提出了一种方法帮助人们判断自己是否超重了,称为身体质量指数,简称BMI。BMI是一个数值,由你多高和多重来定。每个人都有自己的BMI值,你的BMI值可以告诉你是否超重或肥胖,或者是太瘦,还是刚刚好。大部分人一天需要2000卡路里,如果超重,那你需要减少每天所摄入的卡路里来减重。如果太瘦,那么,你可能需要摄入更多的卡路里。

BMI公式

若想知道自己的BMI值，你需要：

1. 称重并记录下你的体重；
2. 量出你高多少英寸（身高）并记录；
3. 使用以下公式：

$$体重 \div (身高 \times 身高) \times 703$$

4. 如果你在20岁以下，那其他因素也将影响你的BMI值。到美国政府的疾病控制与防治中心的网站（apps.nccd.cdc.gov/dnpabmi）上，你可以输入你的度量值并计算出你的BMI值。

15
小　结

现在所有的饭都做完了，拉马尔对于自己能把这顿饭完成得这么好感到非常骄傲——对在厨房中完成的数学及所有的这一切感到骄傲。现在来看一看你是否记得包含在这餐中所有的数学问题。

1. 你想买5袋薯条，每袋薯条1.99美元。总共要花多少钱？

2. 当去商店购物时你带了10美元，要买下列物品：

   牛奶：1.88美元
   面包：0.99美元
   花生酱：3.87美元
   果酱：2.89美元

估计一下你带的钱是否够用。若你想买2条面包呢？带的钱还够用吗？

3. 你发现有牛奶的优惠券。使用优惠券则有50%的折扣。现在你买牛奶要花多少钱？

4. 填写下面的空格：

   5量杯=_____盎司
   16量杯=_____盎司
   6夸脱=_____加仑

5. 一个食谱中需要12量杯的面粉。你想根据食谱做两份，需要多少面粉？

6. 你一次只能放12块饼干进烤箱。如果想做60块饼干，需要做几次？

7. 如果烘焙每批饼干需要花12分钟，那需要多久才能烘焙完60块饼干？

8. 如果每包薯条上注明每份中有10克脂肪，这10克脂肪等于多少卡路里？

9. 将每天所吃的全部卡路里加起来，你发现自己吃了3400的卡路里。若每天都这样吃，那你是会变胖还是变瘦？

参考答案

1.

1.75 × 3 = 5.25
3.75 × 2 = 7.5
3/4 × 2.50 ≈ 1.90

拉马尔有足够多的钱吗？是的，他需要54.90美元，而他有55美元
他的钱有剩余吗？是的
如果有剩余，剩多少？剩余10美分
如果没有剩余，那他还需要多少钱？他有足够多的钱

2.

项目	单价	你的估计/美元	累计/美元
2磅通心粉	0.99美元/磅	2.00	2.00
1磅奶酪	3.69美元/磅	4.00	6.00
1夸脱牛奶	1.53美元/夸脱	2.00	8.00
3磅碎牛肉	1.75美元/磅	6.00	14.00
3个西红柿（3/4磅）	2.50美元/磅	2.00	16.00
1瓶番茄酱	1.87美元/瓶	2.00	18.00
1罐芥末	0.99美元/罐	1.00	19.00
1打汉堡包	3.75美元/6个	8.00	27.00
5磅面粉	3.10美元/5磅	4.00	31.00
5磅糖	2.95美元/5磅	3.00	34.00
烘焙巧克力	4.75美元/包	5.00	39.00
鸡蛋	2.99美元/打	3.00	42.00
发酵粉	3.19美元/罐	4.00	46.00
盐	2.10美元/箱	3.00	49.00
沙拉酱	2.75美元/瓶	3.00	52.00
食用油	3.25美元/瓶	4.00	56.00
1磅黄油	2.80美元/磅	3.00	59.00
细砂糖	2.15美元/包	3.00	62.00
香蕉	0.79美元/把	1.00	63.00
草莓	3.89美元/包	4.00	67.00
辣椒	0.33美元/个	1.00	68.00

拉马尔的钱够吗？不够，通过四舍五入，并加上水果沙拉，拉马尔现在需要68美元

3.

2.80美元 - (2.80美元 ÷ 4) = 2.1美元

20%指100个中有20个。100是20的多少倍？5倍

你的答案是多少？ 3.10美元 - (3.10美元 ÷ 5) = 2.48美元

如果拉马尔使用所有的3张优惠券，他一共可以节省多少钱？他一共节省 (2.82美元 - 2.12美元) + (2.10美元 - 1.05美元) + (3.10美元 - 2.48美元) = 2.37美元

4.

1. 1杯
2. 12盎司
3. 1/2棒, 4汤匙
4. 2杯
5. 4块
6. 9汤匙
7. 8杯

5.

1. 数字乘以3
2. 数字除以3

6.

4人份	+4的1/2份	=6人份
1包通心粉	+1/2包	=1 1/2包通心粉
2个鸡蛋	+1个鸡蛋	=3个鸡蛋
2杯牛奶	+1杯	=3杯
2汤匙融化的黄油	+1汤匙	=3汤匙
2 1/2杯切碎的奶酪	+1 1/4杯	=3 3/4杯

4人份	×3	=12人份
1包通心粉	×3	=3包通心粉
2个鸡蛋	×3	=6个鸡蛋
2杯牛奶	×3	=6杯
2汤匙融化的黄油	×3	=6汤匙
2 1/2杯切碎的奶酪	×3	=7 1/2杯

7.

原蛋糕食谱	翻3倍
1杯面粉	3杯
1杯糖	3杯
3块方形的烘烤巧克力	9块
1茶匙泡打粉	3汤匙
3/4茶匙苏打粉	2 1/4汤匙
1/2汤匙盐	1 1/2汤匙
1/2杯牛奶	1 1/2杯
1/4杯植物油	3/4杯
1个鸡蛋	3个鸡蛋
1汤匙香草精	3汤匙
1/2杯开水	1 1/2杯
1杯中筋面粉	3杯

原糖霜食谱	两倍糖霜食谱
1 1/2杯黄油	3杯
1杯可可粉	2杯
5杯细砂糖	10杯
1/2杯牛奶	1杯
2茶匙香草精	4汤匙

拉马尔需要多少磅黄油制作糖霜？6磅

8.

如果他想让每个人都各有3个汉堡包，那么需要制作几批次？18个汉堡除以每批次4个汉堡=4 1/2批次

如果每批次需要8分钟来制作，那么他制作3批次需要多少时间？

8分钟/炉×3炉=24分钟

如果他要做18个汉堡使得每个人都有3个，一共需要多少时间？18/4×8= 36分钟

	几批？	烹饪时间
你想要烤8片吐司面包，但是烤面包机一次只能放2片面包。烤每一批需要3分钟	4批	12分钟
你烤饼干需要烤12分钟。食谱是36个饼干，但烤盘内一次只能容纳12个饼干	3批	36分钟

9.

如果拉马尔一次做这些事情中的一件，需要多少时间？164分钟，或者2小时44分钟

他能在客人到来之前做完吗？不能，他只有2小时30分钟

如果拉马尔先烘焙蛋糕，蛋糕一出炉就把奶酪通心粉放进烤箱，那么通心粉和奶酪什么时候可以出炉？35分钟+45分钟=80分钟以后

当蛋糕和通心粉在烘焙时，拉马尔和乔治有时间做他们需要做的别的事情吗？别的任务需要84分钟完成，所以他们刚好有时间

10.

	多少杯？
5瓶10盎司的苏打水=	6 1/4杯
2加仑的蔓越莓果汁=	32杯
1/2加仑的苹果汁=	8杯
合计	46 1/4杯

11.

1. 如果他的朋友只吃1份餐，拉马尔的餐里有多少卡路里？891卡路里
2. 如果他们吃2份餐，有多少卡路里呢？1782卡路里
3. 1份餐里有多少糖（单位：克）？和下表所示的推荐每日需求量怎么比较？76克。1份餐中比推荐的每日需求量多64克
4. 钠是盐的别称。1份餐中有多少盐（单位：毫克）？和下表所示的推荐每

日需求量怎么比较？1305毫克。1份餐中比推荐的每日需求量少1005卡路里

    5. 你认为这份餐的卡路里、脂肪、糖和盐的健康度如何？这份餐可以更健康化，因为它含有大量脂肪、糖和盐

12.

1. >
2. <
3. >
4. <

    你认为这些食物的健康度如何？想要健康饮食的人能吃很多吗？为什么可以或不可以？这些食物不健康，因为它们没有包含大量的好营养素，却有一些糖、盐和脂肪。人们应该节制这些食物的食用量

13.

1. 1份奶酪通心粉中有多少卡路里来自脂肪？6克脂肪 × 9卡路里/克 = 54卡路里
2. 巧克力蛋糕中有多少卡路里来自糖？32克糖 × 4卡路里/克 = 128卡路里

看第12节的食物表，回答下列问题：

3. 1份饼干中有多少卡路里来自蛋白质？1克蛋白质 × 4卡路里/克 = 4卡路里
4. 1份沙拉酱中有多少卡路里来自脂肪？6克脂肪 × 4卡路里/克 = 54卡路里

15.

1. 1.99美元 × 5 = 9.95美元
2. 你有足够的钱。如果你买两个面包，钱不够
3. 1.88美元 × 0.5 = 0.94美元
4. 40, 8, 1 1/2
5. 1 1/2 杯
6. 60 ÷ 12 = 5 批次
7. 12分钟 × 5 = 60分钟
8. 10克脂肪 × 每克9卡路里 = 90卡路里
9. 你将会增重，因为你每天只需要2000左右卡路里

INTRODUCTION

How would you define math? It's not as easy as you might think. We know math has to do with numbers. We often think of it as a part, if not the basis, for the sciences, especially natural science, engineering, and medicine. When we think of math, most of us imagine equations and blackboards, formulas and textbooks.

But math is actually far bigger than that. Think about examples like Polykleitos, the fifth-century Greek sculptor, who used math to sculpt the "perfect" male nude. Or remember Leonardo da Vinci? He used geometry—what he called "golden rectangles," rectangles whose dimensions were visually pleasing—to create his famous *Mona Lisa*.

Math and art? Yes, exactly! Mathematics is essential to disciplines as diverse as medicine and the fine arts. Counting, calculation, measurement, and the study of shapes and the motions of physical objects: all these are woven into music and games, science and architecture. In fact, math developed out of everyday necessity, as a way to talk about the world around us. Math gives us a way to perceive the real world—and then allows us to manipulate the world in practical ways.

For example, as soon as two people come together to build something, they need a language to talk about the materials they'll be working with and the object that they would like to build. Imagine trying to build something—anything—without a ruler, without any way of telling someone else a measurement, or even without being able to communicate what the thing will look like when it's done!

The truth is: We use math every day, even when we don't realize that we are. We use it when we go shopping, when we play sports, when we look at the clock, when we travel, when we run a business, and even when we cook. Whether we realize it or not, we use it in countless other ordinary activities as well. Math is pretty much a 24/7 activity!

And yet lots of us think we hate math. We imagine math as the practice of dusty, old college professors writing out calculations endlessly. We have this idea in our heads that math has nothing to do with real life, and we tell ourselves that it's something we don't need to worry about outside of math class, out there in the real world.

But here's the reality: Math helps us do better in many areas of life. Adults who don't understand basic math applications run into lots of problems. The Federal Reserve, for example, found that people who went bankrupt had an average of one and a half times more debt than their income—in other words, if they were making $24,000 per year, they had an average debt of $36,000. There's a basic subtraction problem there that should have told them they were in trouble long before they had to file for bankruptcy!

As an adult, your career—whatever it is—will depend in part on your ability to calculate mathematically. Without math skills, you won't be able to become a scientist or a nurse, an engineer or a computer specialist. You won't be able to get a business degree—or work as a waitress, a construction worker, or at a checkout counter.

Every kind of sport requires math too. From scoring to strategy, you need to understand math—so whether you want to watch a football game on television or become a first-class athlete yourself, math skills will improve your experience.

And then there's the world of computers. All businesses today—from farmers to factories, from restaurants to hair salons—have at least one computer. Gigabytes, data, spreadsheets, and programming all require math comprehension. Sure, there are a lot of automated math functions you can use on your computer, but you need to be able to understand how to use them, and you need to be able to understand the results.

This kind of math is a skill we realize we need only when we are in a situation where we are required to do a quick calculation. Then we sometimes end up scratching our heads, not quite sure how to apply the math we learned in school to the real-life scenario. The books in this series will give you practice applying math to real-life situations, so that you can be ahead of the game. They'll get you started—but to learn more, you'll have to pay attention in math class and do your homework. There's no way around that.

But for the rest of your life—pretty much 24/7—you'll be glad you did!

1
GROCERY MATH

Lamar goes to culinary school, where he is learning to be a chef. Today, he is planning a special meal for five of his friends. They will be celebrating a friend's birthday, so Lamar wants to have all his friend's favorite foods: macaroni and cheese, hamburgers, salad, and chocolate cake. Lamar has chosen recipes, he's checked to see what ingredients he already has on hand, and now he has made a grocery list.

Here's what he plans to buy:

1 pound of macaroni
1 pound of cheese
1 quart of milk
3 pounds of ground beef
1 head of lettuce
3 tomatoes
1 jar of ketchup
1 jar of mustard
1 dozen hamburger buns
5 pounds of flour

5 pounds of sugar
1 package of baking chocolate
1 dozen eggs
a can of baking powder
salt
a bottle of salad dressing
a bottle of vegetable oil
a pound of butter
confectioner's sugar

Lamar has been saving his money, and he has $55 he can spend on the meal. When he gets to the grocery store, he discovers the following prices:

macaroni: $.99 per pound
cheese: $3.69 per pound
milk: $1.53 per quart
ground beef: $1.75 per pound
lettuce: $1.15 per head
tomatoes: $2.50 per pound
ketchup: $1.87 per bottle
mustard: $0.99 per bottle
hamburger buns: $3.75 for six
flour: $3.10 for 5 pounds

sugar: $2.95 for 5 pounds
baking chocolate: $4.75 per package
eggs: $2.99 per dozen
baking powder: $3.19 per can
salt: $2.10 per container
salad dressing: $2.75
cooking oil: $3.25 per bottle
butter: $2.80 per pound
confectioner's sugar: $2.15 per bag

How much will all this cost?

To find out, you'll need to add up the costs of each item. First, though, multiply wherever you need to.

If Lamar needs 3 pounds of ground beef, you'll need to multiply the cost for 1 pound by 3:

$$1.75 \times 3 =$$

If Lamar needs a dozen buns, you'll need to multiply the cost of 6 buns by 2:

$$3.75 \times 2 =$$

If he needs 3 tomatoes, he needs to know how much they will weigh. The average tomato weighs about 4 ounces, which means that 4 tomatoes weigh about a pound (16 ounces). Lamar only wants 3 tomatoes, though. So you will need to find the answer to this equation:

$$¾ \times 2.50 =$$

Now you need to add up everything.
Does Lamar have enough money?
Does he have any left over?
If so, how much?
And if not, how much more money does he need?

2
ESTIMATING AT THE GROCERY STORE

Once Lamar gets to the grocery store, he decides he'd like to make a fruit salad too. He picks up some bananas and strawberries. And the peppers look good to him as well; he wants to include them in the tossed salad.

But he's not sure he'll have enough money. The bananas cost 79 cents, the strawberries cost $3.89, and the peppers cost 33 cents each. He doesn't have a calculator with him to add up the amounts like you just did, and it's too many numbers for him to do in his head.

So how can Lamar avoid a nasty surprise when he gets to the checkout counter? He doesn't want to wait in line only to find out that he doesn't have enough money with him to pay for his groceries!

Estimating in your head is the way to have a good idea how much you are spending on groceries. You do that by rounding up to nearest dollar amount.

Fill out the chart on the next page. The first two entries have been done for you. When you're done, you'll be able to see whether Lamar can afford to buy the additional groceries.

ITEM	PRICE PER UNIT	YOUR ESTIMATE	RUNNING TOTAL
2 lbs. macaroni	$0.99 per lb	$2.00	$2.00
1 lb cheese	$3.69 per lb	$4.00	$6.00
1 qt milk	$1.53 per qt		
3 lbs ground beef	$1.75 per lb		
3 tomatoes (3/4 lb)	$2.50 per lb		
bottle of ketchup	$1.87 per bottle		
jar of mustard	$0.99 per jar		
dozen buns	$3.75 for 6		
5 lbs flour	$3.10 for 5 lbs		
5 lbs sugar	$2.95 for 5 lbs		
baking chocolate	$4.75 per package		
eggs	$2.99 per dozen		
baking powder	$3.19 per can		
salt	$2.10 per container		
salad dressing	$2.75 per bottle		
cooking oil	$3.25 per bottle		
1 lb butter	$2.80 per lb		
confectioner's sugar	$2.15 per bag		
bananas	$0.79 per bunch		
strawberries	$3.89 per package		
peppers	$0.33 each		

Does Lamar have enough money?

3
GROCERY SHOPPING WITH COUPONS

While Lamar is shopping, he runs into his friend Kaylee, who is also doing some shopping. Kaylee has some coupons with her that she shares with Lamar. Now he won't have to spend quite so much money when he goes through the line.

Coupons are usually printed in newspapers and magazines. You'll need to cut out the coupons and bring them with you to the store. Sometimes you can find them online too, and then you'll need to print them up. Give your coupons to the cashier when you pay for the item, and she'll take off some money for that item.

How much money will Lamar save if he uses the coupons on the next page?

Percent means out of 100, so 25 percent means 25 out of 100. That's the same thing as one-fourth. (You can remember this, if you think about quarters, which equal 25 cents each—and there are 4 quarters in a dollar.) So if something is 25% off, that means that first you need to find out what one-fourth of the price is. You can do that by dividing the cost of butter by 4—and then subtracting that number from the cost.

Your equation will look like this:

$$\$2.80 - (\$2.80 \div 4) =$$

(Remember that you always do the part of the problem that's inside parentheses first.)

50% is the same thing as 50 out of 100—and that's the same thing as one-half. An easy way to find out how much something will cost if it's 50% off, is to find out how much it would cost if you took away one-half of the price. In this case, if you divide the cost of salt by 2, you'll know how much Lamar will have to spend on salt with the coupon.

20% means 20 out of 100. How many times does 20 go into 100?

Once you know that, you'll know the number to divide the cost of flour by.

And then you subtract that number from the cost.

What's your answer?

There's also another way you can figure out percents: you can use decimals. To do this, you'll need to remember that you have to move the decimal point two places. So 20% = .20. Then you can multiply the cost of flour by .20 (3.10 x .20), and you'll find out what 20% equals. Then you subtract that amount from $3.10 to find out what 20% off equals.

Either way you do it, you'll get the same answer. Use whichever way is easier for you to do in your head.

How much will Lamar save altogether if he uses all 3 coupons?

4 KITCHEN MEASUREMENTS

Now that Lamar has bought all his groceries, he's ready to start preparing the food. He lines up the groceries on the kitchen counter and gets out his recipes. The recipes, though, call for measurements that aren't listed on the food packages. He'll need to be able to convert from one unit of measurement to another in order to be sure he used the right amount of **ingredients**.

If you use the chart on the following page, you should be able to **convert** one kind of measurement into another fairly easily. You will need to multiply or divide to find the correct answers.

1 cup of grated cheese = 8 ounces
3 teaspoons = 1 tablespoon   16 tablespoons = 1 cup
1 cup = 8 ounces = 1/2 pint   2 cups = 16 ounces = 1 pint
4 cups = 32 ounces = 2 pints = 1 quart   2 quarts = 64 ounces = ½ gallon
4 quarts = 128 ounces = 1 gallon   1 square of baking chocolate = 1 ounce
1 stick of butter = 8 tablespoons = ½ cup = 4 ounces

1. Lamar's macaroni and cheese recipe calls for 8 ounces of grated cheese. How many cups will he need?

2. If the recipe needs 1½ cups of milk, how many ounces is that?

3. The recipe asks for ¼ cup of butter. How much of a stick of butter will that be? How many tablespoons will it be?

4. Lamar's recipe for chocolate cake calls for 16 ounces of flour. How many cups is that?

5. The cake needs 4 ounces of baking chocolate. How many squares will that be?

6. If Lamar needs 3 tablespoons of baking powder, how many teaspoons will that be?

7. How many cups will give him 2 quarts of water?

5
FRACTIONS IN THE KITCHEN

Recipes usually tell you how many servings they make. But sometimes you need to make a different number of servings. When that happens, you must adjust the amount of each ingredient in the recipe. Especially when you're making baked goods, you can't just throw in extra of one ingredient and expect the food to turn out right. Recipes for cakes, cookies, and breads are like chemical formulas: you need to keep the amounts exactly right or the cake, cookies, or bread won't rise right. It might turn out flat and hard—or it might rise too much and be full of bubbles.

So if you want to double a recipe you need to multiply each ingredient by 2. And if you want to make half of a recipe, you'll need use exactly half of each ingredient. You'll find some examples on the next page.

If you want to double a recipe that calls for 2½ cups of flour, first you will need to make improper fractions:

$$2 = 2/1$$
$$2½ = 5/2$$

Then multiply them together:

$$2/1 \times 5/2 = 10/2 = 5$$

So now you know you need 5 cups of flour to double the recipe.

This time suppose a recipe calls for 3½ cups of flour, but you want to make only half of the recipe. This time you will multiply by ½. When you turn 3½ into an improper fraction, you get 7/2. So:

$$½ \times 7/2 = 7/4 = 1¾$$

So to make half the recipe, you will need 1¾ cups of flour.

An easy way to get the same answers is to remember these rules:

Double Recipe: To get twice as many servings, multiply the amount of each ingredient by 2.
Half Recipe: To get half the servings, divide the amount of each ingredient by 2.

But remember, if you're working with fractions, you'll need to turn them into improper fractions first before you multiply!

Now see if you can figure out the answers to these questions:

1. What about if a recipe makes 4 servings—but you need 12 servings? Will you need to multiply the numbers in the recipe or divide? What number will you use?

2. What if the recipe makes 15 servings, but you only want to make 5. Will you need to multiply or divide? What number will you use?

6
RECIPE MATH

Lamar is making macaroni and cheese for his friends. He wants to be able to serve 6 people and have enough for everyone to have one helping—but the recipe he's using only makes 4 servings. What will Lamar need to do?

This is a little more tricky. He needs his recipe to make ½ more than it does. So first he'll need to divide each ingredient amount by 2. And then he'll need to add that amount to the original amount.

On the next page, change the recipe so that it will make 6 servings instead of 4. The first ingredient has been done for you.

4 servings	+ ½ of 4	= 6 servings
1 package of macaroni	+ ½ package	= 1½ packages of macaroni
2 eggs		
2 cups milk		
2 tablespoons melted butter		
2½ cups shredded cheese		

On second thought, though, Lamar decides he wants to have enough macaroni and cheese so that everyone can have second helpings. This means he will need 12 servings instead. Can you fill out the chart below, changing the amount of ingredients so that this time he will end up with 12 servings?

4 servings	× 3	= 12 servings
1 package of macaroni	× 3	= 3 packages of macaroni
2 eggs		
2 cups milk		
2 tablespoons melted butter		
2½ cups shredded cheese		

7
MORE RECIPE MATH

Lamar also needs to bake the chocolate cake. He's decided to invite more people to drop by for dessert, so he wants to be sure the cake is big enough to serve 20 people. The recipe he wants to use, though, will only serve 8. How can he figure out how to make a cake big enough for that many people?

He has a couple of options. He could divide 20 by 8, which equals 2 with a remainder of 4. Since 4 is half of 8, this is the same thing as 2½.

$$20 \div 8 = 2½$$

Now he will need to multiply each of the ingredients in his recipe by 2½.

But Lamar wants to make it easier for himself. He decides to triple the recipe instead. This means he will multiply all the measurements by 3 instead. Multiplying by a whole number is easier.

Now he can cut the cake into slightly bigger slices and still have enough for everyone. Or he can have 4 pieces left over to keep for himself!

Fill in the chart on the next page to find out the amounts he'll need for each ingredient if he triples the recipe.

Original Cake Recipe	Tripled
1 cup flour	
1 cup sugar	
3 square baking chocolate	
1 teaspoons baking powder	
¾ teaspoons baking soda	
½ teaspoon salt	
½ cup milk	
¼ cup vegetable oil	
1 egg	
1 teaspoon vanilla extract	
½ cup boiling water	
1 cup all-purpose flour	

Now Lamar will need to be sure he has enough frosting to cover the cake. He probably won't need exactly 3 times as much, but the cake will be bigger, so he's going to need more frosting. He decides to double the frosting recipe. Can you find the amounts he'll need?

Original Frosting Recipe	Doubled
1½ cups butter	
1 cup cocoa	
5 cups confectioner's sugar	
½ cup milk	
2 teaspoons vanilla extract	

How many sticks of butter will Lamar need now for the frosting? Look back to section 6 if you need to.

8
COOKING IN BATCHES

Lamar has a small grill on his deck where he is going to cook the hamburgers. However, the grill will only hold 4 burgers at a time. If Lamar wants to make enough burgers for each person to have 2, he'll need 12 burgers because he's expecting 6 people. This means he'll need to cook a batch at a time.

How many batches will he need?

$$12 \div 4 = 3$$

He'll need to make 3 batches.

If he wants to make enough for everyone to have 3 burgers, how many batches will he need to make?

Now Lamar realizes that it's going to take him a while to make that many batches of burgers. He's going to need to allow time for all the batches to cook.

If it takes 8 minutes for one batch to cook, how long will it take for him to cook 3 batches?

If he makes 18 burgers, enough for everyone to have 3, how long will he need to allow for the cooking time?

PRACTICES WITH BATCHES

Fill out the following chart for more practice.

	How Many Batches?	Cooking Time
You want to make 8 pieces of toast, but your toaster will only hold 2 slices of bread at a time. It takes 3 minutes for each batch to toast.		
You're baking cookies that need 12 minutes to bake. The recipe will make 36 cookies, but your cookie sheet will only hold 12 cookies at a time.		

9
USING TIME IN THE KITCHEN

Lamar's friend George stops and says he will help Lamar get the food ready. Lamar asks George to cut up the tomatoes for the salad—but George works so slowly that Lamar is worried the tomatoes will never be cut. Lamar takes a deep breath and looks at the clock to figure out how much time he has.

It's 4:00PM now, and his friends will be arriving at 6:30PM. That means he still has two and a half more hours to get ready.

The macaroni and cheese is made and it's ready to go in the oven. The cake is also made and ready to be baked, but he can't fit both the cake and the macaroni in the oven at the same time (and besides, they need different temperatures).

Here are the things Lamar still needs to do, with how much time each will take.

Task to Be Done	Time It Will Take
Bake the cake	35 minutes
Cool the cake	10 minutes
Make the frosting	10 minutes
Ice the cake	10 minutes
Bake the macaroni and cheese	45 minutes
Make the tossed salad	10 minutes
Make the fruit salad	10 minutes
Set the table	10 minutes
Grill the burgers	24 minutes

If Lamar does each one of these things, one at a time, how much time will it take?

Will he be done before his guests arrive?

The good news is—while things that need to go in the oven are baking, Lamar and George can be doing other tasks!

Lamar decides it makes sense to grill the hamburgers last, so they'll be as hot as possible. He decides to bake the cake first, so that it will have time to cool before he frosts it.

So if Lamar bakes the cake first, then puts the macaroni and cheese into the oven as soon as the cake comes out, what time will the macaroni and cheese be ready to come out of the oven?

While the cake and the macaroni are baking, will Lamar and George have time to do the other things they need to do?

10
LIQUID MATH

Lamar's friend Rashelle drops by and offers to help him and George prepare the meal. She's brought a few other things with her to add to the meal: a package of oatmeal cookies because Kaylee doesn't like chocolate and several bottles of soda and juice. She and Lamar decide to make punch by combining the soda and juice.

Rashelle brought 5 ten-ounce bottles of ginger ale, 2 gallon jugs of cranberry juice, and a half-gallon of apple juice. If she combines these altogether, how many cup servings will she have?

Look back to Section 6 to fill out this chart.

	How Many Cups?
5 ten-ounce bottles of soda =	
2 gallons of cranberry juice =	
½ gallon of apple juice =	
TOTAL	

WHAT ABOUT METRIC?

Metric measurements are another way to measure liquids.
1 gallon = 3.8 liters
1 quart = .9 liters
1 cup = .2 liters

50

11 NUTRITION MATH

When Lamar's friend Ritesh arrives, he wants to find out how nutritious the meal will be. Ritesh is also going to culinary school, but he's focusing more on nutrition. He explains to his friends that everybody needs a balance of different nutrients in their diets.

Rashelle wants to know how fattening their meal will be, and if the foods are healthy ones. Ritesh explains that calories, fat, salt, and sugar are all important factors in how healthy a particular food is. Unsaturated fat is better for you than saturated, he says. He pulls out his laptop and goes to a website that lets him type in different foods to find out their nutritional values.

The results he gets are shown below. Most of these numbers are given in grams and milligrams. One gram weighs about as much as paperclip. A milligram is lighter, and weighs about the same as a snowflake.

1 Serving	Calories	Sugar	Fat	Saturated Fat	Sodium (Salt)
macaroni & cheese	200	1 g	6 g	2 g	1027 mg
hamburger	230	0 g	15 g	6 g	64 mg
tossed salad	25	0 g	0 g	0 g	0 mg
fruit salad	50	7 g	0 g	0 g	0 mg
punch	150	36 g	0 g	0 g	0 mg
chocolate cake	236	32 g	10 g	3 g	214 mg

1. How many calories are in Lamar's meal, if his friends only eat one serving?
2. What about if they eat two servings?
3. How much sugar (in grams) is in one serving of the meal? How does this compare to the recommended daily requirement shown in the chart below?

4. Sodium is another name for salt. How much salt (in milligrams) is in one serving of this meal?
5. How healthy do you think this meal is when it comes to calories, fat, sugar, and salt?

Daily Diet Reccomendations:

	Men	Women	Children 4-8	Boys 9-13	Boys 14-18	Girls 9-13	Girls 14-18
Fat	20%–35% of daily calories	20%–35% of daily calories	25%–35% of daily calories	25%–35% of daily calories	25%–35% of daily calories	25%–35% of daily calories	25%–35% of daily calories
Saturated Fat	less than 10% of daily calories	less than 10% of daily calories	less than 10% of daily calories	less than 10% of daily calories	less than 10% of daily calories	less than 10% of daily calories	less than 10% of daily calories
Sodium	less than 2300 mg	less than 2300 mg	less than 1900 mg	less than 2200 mg	less than 2300 mg	less than 2200 mg	less than 2300 mg
Sugar	37.5 grams	25 grams	12 grams	12 grams	12 grams	12 grams	12 grams

Sources: U.S. Food and Drug Administration and the American Heart Association.

12
GREATER THAN AND LESS THAN: FOOD LABEL MATH

Rashelle points out that Ritesh left out the salad dressing when he was looking up things online. Before Ritesh can open up his laptop again, Lamar tells his friends that a lot of foods come with labels that tell you how many nutrients it has and if it's healthy or not. Foods like salad dressing that come in bottles or other foods in boxes or bags have labels that give lots of nutrition information.

You'll find food labels on just about every food with a package. Look for a black and white chart with a whole bunch of numbers. You need 100 percent of every nutrient every day to be at your healthiest. Daily values tell you how close you are to getting 100 percent of each nutrient.

Nutrition Facts

Serving Size 2 Tbsp (30g)
Servings Per Container about 12
Calories 70
  Calories from Fat 50
*Percent Daily Values are based on a 2,000 calorie diet.

Amount / Serving	% Daily Value*	Amount / Serving	% Daily Value*
Total Fat 6g	9%	**Total Carbohydrate** 3g	1%
Saturated Fat 3.5g	18%	Dietary Fiber 0g	0%
Trans Fat 0g		Sugars 2g	
Cholesterol 20mg	7%	**Protein** 1g	
Sodium 180mg	7%		
Vitamin A 6% • Vitamin C 2%		Calcium 4% • Iron 0%	

Compare the salad dressing food label above with the food label below from the cookies that Rashelle brought. Insert either < or > into each blank below:

1. Which food has more fat per serving?

salad dressing _____ cookies

2. Which food has more calories per serving?

salad dressing _____ cookies

3. Which one has more salt?

salad dressing _____ cookies

4. Which one has more sugar per serving?

salad dressing _____ cookies

How healthy do you think these foods are? Should people who want a healthy diet eat a lot of them? Why or why not?

Nutrition Facts

Serving Size 1 Cookie (17g)
Servings Per Container About 12
Calories 80
  Calories from Fat 40
*Percent Daily Values are based on a 2,000 calorie diet.

Amount / Serving	% Daily Value*	Amount / Serving	% Daily Value*
Total Fat 4.5g	7%	**Total Carbohydrate** 7g	2%
Saturated Fat 2.5g	13%	Dietary Fiber 1g	4%
Trans Fat 0g		Sugars 4g	
Cholesterol 5mg	2%	**Protein** 1g	
Sodium 20mg	1%		
Vitamin A 0% • Vitamin C 0%		Calcium 0% • Iron 2%	

13
COUNTING CALORIES

Ritesh explains to his friends that there's a connection between different kinds of nutrients and the number of calories in foods.

"Our bodies need different nutrients from the food we eat," Ritesh says. "We need a balance of protein, carbohydrates, and fat. We need protein to keep our muscles healthy. We need carbohydrates for energy. And we need healthy fat to help us absorb vitamins and for energy."

We get protein from meat and dairy products, like cheese and milk. Carbohydrates are in breads, desserts, fruits, and vegetables. Fat is found in meat and dairy products, as well as in many snacks. Each kind of nutrient has a different number of calories.

Protein, for example, is a nutrient that has 4 calories in every gram. If you were eating some cheese with 7 grams of protein, you would be eating 28 calories that came from protein.

7 grams protein x 4 calories per gram = 28 calories

Other calories in the cheese come from other nutrients.

Carbohydrates also have four calories per gram. Soda, for example, has a lot of sugar, which is a carbohydrate. One can of soda usually has around 40 grams of sugar. In all, there are 160 calories from that sugar.

40 grams sugar x 4 calories per gram = 160 calories

Soda doesn't really have anything else in it, so all the calories in it come from sugar.

Fat has the most calories per gram. Fat has 9 calories for every gram. Half an avocado has about eleven grams of healthy, unsaturated fat. When you do the math, you see half an avocado has 99 calories.

11 grams of fat x 9 calories per gram = 99 calories

Look back on Section 11 to answer these questions:

1. How many calories come from fat in a serving of macaroni and cheese?
2. How many calories come from sugar in chocolate cake?

Now look at the food labels on Section 12 to answer these questions:

3. How many calories come from protein in one cookie?
4. How many calories come from fat in serving of salad dressing?

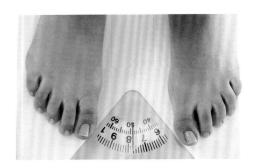

14 HOW MANY CALORIES DO YOU NEED?

While Lamar and his friends eat the meal he prepared, they laugh and have a good time. After dinner, Kaylee and George offer to clean up the kitchen, while the others hang out in the living room.

As they are putting the dishes in the dishwasher, they talk some more about calories. George tells Kaylee he's worried he may be overweight and wonders if he should go on a diet. Kaylee explains to him that you can't just stand on the scales to tell whether you weigh too much. Everyone's body is different. Some people have heavier bones. Some have more muscle. Some are tall, short, and everything in between.

Doctors have come up with a way to tell if people weigh too much. They call it the Body Mass Index, or BMI. BMI depends on how tall you are and how much you weigh. It's a number. Everyone has a BMI. Your BMI can tell you if you're overweight or obese—or if you're underweight or just right. Most people need about 2,000 calories a day. If you're overweight, then you should cut back on your calories to lose weight. If you're underweight, though, you might need to eat more calories.

BMI Formula

If you want to figure out your own BMI, here's what you need to do:

1. Weigh yourself and write it down.
2. Measure how tall you are in inches (height) and write it down.
3. Use this equation.

Weight ÷ (height x height) x 703

4. If you're under age 20, though, other factors will affect your BMI. Go to the U.S. government's Centers for Disease Control and Prevention (CDC) website at apps.nccd.cdc.gov/dnpabmi. You'll be able to plug in your measurements there to calculate your BMI.

15
PUTTING IT ALL TOGETHER

Now that all the cooking is done, Lamar feels pretty proud of himself that he was able to pull off the meal so well—math and all.

See if you can remember all the math that went into this meal.

1. A bag of potato chips cost $1.99. You want to buy 5 bags. How much will they cost altogether?

2. You have $10 with you when you go to the store for groceries. You need to buy these things:

   milk: $1.88
   bread: $.99
   peanut butter: $3.87
   jelly: $2.89

   Estimate to find out if you have enough money with you. What about if you buy 2 loaves of bread? Will you still have enough?

3. You discover you have a coupon with you for the milk. The coupon is for 50% off. How much will the milk cost now?

4. Fill in the blanks below:

   5 cups = _____ ounces
   16 cups = _____ pints
   6 quarts = _____ gallons

5. A recipe calls for ¾ cup flour. You want to double the recipe. How much flour will you need now?

6. You can only fit 12 cookies at a time into the oven. You want to make 60 cookies. How many batches will you need to make?

7. If each batch of cookies takes 12 minutes to bake, how long will you need to bake all 60 cookies?

8. If a package of potato chips says it has 10 grams of fat per serving, how many calories do those 10 grams equal?

9. You add up all the calories you eat in a day, and find you ate 3,400 calories. If you eat like that every day, will you probably gain weight or lose weight?

ANSWERS

1.

1.75 x 3 = 5.25
3.75 x 2 = 7.5
¾ x 2.50 ≈ 1.90

Does Lamar have enough money? Yes. He needs $54.90, and he has $55.
Does he have any left over? Yes.
If so, how much? He has ten cents left.
And if not, how much more money does he need? He has enough money.

2.

ITEM	PRICE PER UNIT	YOUR ESTIMATE	RUNNING TOTAL
2 lbs. macaroni	$0.99 per lb	$2.00	$2.00
1 lb cheese	$3.69 per lb	$4.00	$6.00
1 qt milk	$1.53 per qt	$2.00	$8.00
3 lbs ground beef	$1.75 per lb	$6.00	$14.00
3 tomatoes (3/4 lb)	$2.50 per lb	$2.00	$16.00
bottle of ketchup	$1.87 per bottle	$2.00	$18.00
jar of mustard	$0.99 per jar	$1.00	$19.00
dozen buns	$3.75 for 6	$8.00	$27.00
5 lbs flour	$3.10 for 5 lbs	$4.00	$31.00
5 lbs sugar	$2.95 for 5 lbs	$3.00	$34.00
baking chocolate	$4.75 per package	$5.00	$39.00
eggs	$2.99 per dozen	$3.00	$42.00
baking powder	$3.19 per can	$4.00	$46.00
salt	$2.10 per container	$3.00	$49.00
salad dressing	$2.75 per bottle	$3.00	$52.00
cooking oil	$3.25 per bottle	$4.00	$56.00
1 lb butter	$2.80 per lb	$3.00	$59.00
confectioner's sugar	$2.15 per bag	$3.00	$62.00
bananas	$0.79 per bunch	$1.00	$63.00
strawberries	$3.89 per package	$4.00	$67.00
peppers	$0.33 each	$1.00	$68.00

Does Lamar have enough money? No. By rounding up to the nearest dollar, and adding in the fruit salad, Lamar needs $68 now.

3.

$2.80 − ($2.80 ÷ 4) = $2.1

20% means 20 out of 100. How many times does 20 go into 100? 5 times.

What's your answer? $3.10 − ($3.10 ÷ 5) = $2.48

How much will Lamar save altogether if he uses all 3 coupons? He's saving ($2.82 − $2.12) + ($2.10 − $1.05) + ($3.10 − $2.48) = $2.37 in savings.

4.

1. 1 cup
2. 12 ounces
3. ½ stick, 4 tablespoons
4. 2 cups
5. 4 squares
6. 9 teaspoons
7. 8 cups

5.

1. Multiply the numbers by 3.
2. Divide the numbers by 3.

6.

4 servings	+ ½ of 4	= 6 servings
1 package of macaroni	+ ½ package	= 1½ packages of macaroni
2 eggs	+ 1 egg	= 3 eggs
2 cups milk	+ 1 cup	= 3 cups
2 tablespoons melted butter	+ 1 tablespoon	= 3 tablespoons
2½ cups shredded cheese	+ 1 ¼ cup	= 3 ¾ cups

4 servings	x 3	= 12 servings
1 package of macaroni	x 3	= 3 packages of macaroni
2 eggs	x 3	= 6 eggs
2 cups milk	x 3	= 6 cups
2 tablespoons melted butter	x 3	= 6 tablespoons
2½ cups shredded cheese	x 3	= 7 ½ cups

7.

Original Cake Recipe	Tripled
1 cup flour	3 cups
1 cup sugar	3 cups
3 square baking chocolate	9 squares

1 teaspoons baking powder	3 teaspoons
3/4 teaspoons baking soda	2 ¼ teaspoons
½ teaspoon salt	1 ½ teaspoons
½ cup milk	1 ½ cups
¼ cup vegetable oil	¾ cup
1 egg	3 eggs
1 teaspoon vanilla extract	3 teaspoons
½ cup boiling water	1 ½ cups
1 cup all-purpose flour	3 cups

Original Frosting Recipe	**Doubled**
1½ cups butter	3 cups
1 cup cocoa	2 cups
5 cups confectioner's sugar	10 cups
½ cup milk	1 cup
2 teaspoons vanilla extract	4 teaspoons

How many sticks of butter will Lamar need now for the frosting? 6 sticks.

8.

If he wants to make enough for everyone to have 3 burgers, how many batches will he need to make? 18 burgers divided by 4 burger per batch= 4 ½ batches

If it takes 8 minutes for one batch to cook, how long will it take for him to cook 3 batches? 8 minutes x 3 batches= 24 minutes

If he makes 18 burgers, enough for everyone to have 3, how long will he need to allow for the cooking time? $18/4 \times 8 = 36$ minutes

	How Many Batches?	**Cooking Time**
You want to make 8 pieces of toast, but your toaster will only hold 2 slices of bread at a time. It takes 3 minutes for each batch to toast.	4 batches	12 minutes

You're baking cookies that need 12 minutes to bake. The recipe will make 36 cookies, but your cookie sheet will only hold 12 cookies at a time.	3 batches	36 minutes

9.

If Lamar does each one of these things, one at a time, how much time will it take? 164 minutes, or 2 hours and 44 minutes.

Will he be done before his guests arrive? Not quite. He only has 2 hours and 30 minutes.

So if Lamar bakes the cake first, then puts the macaroni and cheese into the oven as soon as the cake comes out, what time will the macaroni and cheese be ready to come out of the oven? 35 minutes + 45 minutes= 80 minutes from now.

While the cake and the macaroni are baking, will Lamar and George have time to do the other things they need to do? The other tasks will take 84 minutes to do, so they will just about have time.

10.

	How Many Cups?
5 ten-ounce bottles of soda =	6 ¼ cups
2 gallons of cranberry juice =	32 cups
½ gallon of apple juice =	8 cups
TOTAL	46 ¼ cups

11.

1. How many calories are in Lamar's meal, if his friends only eat one serving? 891 calories
2. What about if they eat two servings? 1782 calories.
3. How much sugar (in grams) is in one serving of the meal? How does this compare to the recommended daily requirement shown in the chart below? 76 grams. One serving has 64 more grams of sugar than the daily recommended value.
4. Sodium is another name for salt. How much salt (in milligrams) is in one serving of this meal? How does this compare to to the recommended daily requirement show in the chart below? 1305 milligrams. One serving has 1005 fewer calories than the recommended requirement.

5. How healthy do you think this meal is when it comes to calories, fat, sugar, and salt? This meal could be a lot healthier, because it has a lot of fat, sugar, and salt.

12.

1. >
2. <
3. >
4. <

How healthy do you think these foods are? Should people who want a healthy diet eat a lot of them? Why or why not? These foods aren't that healthy because they don't have a lot of good nutrients in them, but they do have some sugar, salt, and fat. People should limit how much they eat of them.

13.

1. How many calories come from fat in a serving of macaroni and cheese? 6 grams of fat x 9 calories per gram= 54 calories
2. How many calories come from sugar in chocolate cake? 32 grams of sugar x 4 calories per gram= 128 calories

Now look at the food labels on section 12 to answer these questions:

3. How many calories come from protein in one cookie? 1 gram of protein x 4 calories per gram= 4 calories
4. How many calories come from fat in serving of salad dressing? 6 grams of fat x 9 calories per gram= 54 calories.

15.

1. $1.99 x 5 = $9.95
2. You do have enough money. You won't have enough if you buy two loaves of bread.
3. $1.88 x .5 = $.94
4. 40, 8, 1 ½
5. 1 ½ cups
6. 60 ÷ 12 = 5 batches
7. 12 minutes x 5 = 60 minutes
8. 10 grams of fat x 9 calories per gram = 90 calories
9. You will gain weight, because you only need around 2,000 calories a day.

本书由中国科学院数学与系统科学研究院资助出版

数学 24/7

购物中的数学

〔美〕海伦·汤普森 著

赵 晶 邵伟文 译

科学出版社

北京

图字：01-2015-5627号

内 容 简 介

购物中的数学是"数学生活"系列之一，内容涉及购物的预算、消费税、付款方式、物品价格及成本、在线购物、优惠和折扣等方面，同时比较了不同付款方式的优缺点，不同优惠和折扣之间的比较和计算，让青少年在学校学到的数学知识应用到与购物有关的多个方面中，让青少年进一步了解数学在日常生活中是如何运用的。

本书适合作为中小学生的课外辅导书，也可作为中小学生的兴趣读物。

Copyright © 2014 by Mason Crest, an imprint of National Highlights, Inc. All rights reserved. No part of this publication may be reproduced or transmitted in any form or by any means, electronic or mechanical, including photocopying, recording, taping or any information storage and retrieval system, without permission from the publisher.
The simplified Chinese translation rights arranged through Rightol Media.
（本书中文简体版权经由锐拓传媒取得Email:copyright@rightol.com）

图书在版编目（CIP）数据

购物中的数学/（美）海伦·汤普森（Helen Thompson）著；赵晶，邵伟文译.—北京:科学出版社,2018.5
（数学生活）
书名原文：Shopping Math
ISBN 978-7-03-056358-3

Ⅰ.①购… Ⅱ.①海…②赵…③邵… Ⅲ.①数学-青少年读物 Ⅳ.①01-49
中国版本图书馆CIP数据核字（2018）第011939号

责任编辑:胡庆家 / 责任校对:邹慧卿
责任印制:肖 兴 / 封面设计:陈 敬

科 学 出 版 社 出版
北京东黄城根北街16号
邮政编码：100717
http://www.sciencep.com

北京汇瑞嘉合文化发展有限公司 印刷
科学出版社发行 各地新华书店经销

*

2018年5月第 一 版 开本:889×1194 1/16
2018年5月第一次印刷 印张:4 1/4
字数:70 000

定价：98.00元（含2册）
（如有印装质量问题，我社负责调换）

引　　言

你会如何定义数学？它也许不是你想象的那样简单。我们都知道数学和数字有关。我们常常认为它是科学，尤其是自然科学、工程和医药学的一部分，甚至是基础部分。谈及数学，大多数人会想到方程和黑板、公式和课本。

但其实数学远不止这些。例如，在公元前5世纪，古希腊雕刻家波留克列特斯曾经用数学雕刻出了"完美"的人体像。又例如，还记得列昂纳多·达·芬奇吗？他曾使用有着赏心悦目的尺寸的几何矩形——他称之为"黄金矩形"，创作出了著名的画作——蒙娜丽莎。

数学和艺术？是的！数学对包括医药和美术在内的诸多学科都至关重要。计数、计算、测量、对图形和物理运动的研究，这些都被融入到音乐与游戏、科学与建筑之中。事实上，作为一种描述我们周围世界的方式，数学形成于日常生活的需要。数学给我们提供了一种去理解真实世界的方法——继而用切实可行的途径来控制世界。

例如，当两个人合作建造一样东西时，他们肯定需要一种语言来讨论将要使用的材料和要建造的对象。但如果他们建造的过程中没有用到一个标尺，也不用任何方式告诉对方尺寸，甚至他们不能互相交流，那他们建造出来的东西会是什么样的呢？

事实上，即便没有察觉到，但我们确实每天都在使用数学。当我们购物、运动、查看时间、外出旅行、出差办事，甚至烹饪时都用到了数学。无论有没有意识到，我们在数不清的日常活动中用着数学。数学几乎每时每刻都在发生。

很多人都觉得自己讨厌数学。在我们的想象中，数学就是枯燥乏味的老教授做着无穷无尽的计算。我们会认为数学和实际生活没有关系；离开了数学课堂，在真实世界里我们再不用考虑与数学有关的事情了。

然而事实却是数学使我们生活各方面变得更好。不懂得基本的数学应用的人会遇到很多问题。例如，美联储发现，那些破产的人的负债是他们所得收入的1.5倍左右——换句话说，假设他们年收入是24000美元，那么平均负债是36000美元。懂得基本的减法，会使他们提前意识到风险从而避免破产。

作为一个成年人，无论你的职业是什么，都会或多或少地依赖于你的数学计算能力。没有数学技巧，你就无法成为科学家、护士、工程师或者计算机专家，就无法得到商学院学位，就无法成为一名服务生、一位建造师或收银员。

体育运动也需要数学。从得分到战术，都需要你理解数学——所以无论你是想在电视上看一场足球比赛，还是想在赛场上成为一流的运动员，数学技巧都会给你带来更好的体验。

还有计算机的使用。从农庄到工厂、从餐馆到理发店，如今所有的商家都至少拥有一台电脑。千兆字节、数据、电子表格、程序设计，这些都要求你对数学有一定的理解能力。当然，电脑会提供很多自动运算的数学函数，但你还得知道如何使用这些函数，你得理解电脑运行结果的含义。

这类数学是一种技能，但我们总是在需要做快速计算时才会意识到自己需要这种技能。于是，有时我们会抓耳挠腮，不知道如何将学校里学的数学应用在实际生活中。这套丛书将助你一马当先，让你提前练习数学在各种生活情境里的运用。这套丛书将会带你入门——但如果想掌握更多，你必须专心上数学课，认真完成作业，除此之外再无捷径。

但是，付出的这些努力会在之后的生活里——几乎每时每刻（24/7）——让你受益匪浅！

目　　录

引言
1. 使用购物单　　　　　　　　　　　　　　　　1
2. 预算　　　　　　　　　　　　　　　　　　　3
3. 停车　　　　　　　　　　　　　　　　　　　5
4. 消费税　　　　　　　　　　　　　　　　　　7
5. 现金消费　　　　　　　　　　　　　　　　　9
6. 使用借记卡　　　　　　　　　　　　　　　　11
7. 使用信用卡　　　　　　　　　　　　　　　　13
8. 衣服的尺码　　　　　　　　　　　　　　　　15
9. 衣服的性价比　　　　　　　　　　　　　　　17
10. 销售和折扣　　　　　　　　　　　　　　　 19
11. 促销：买一送一　　　　　　　　　　　　　 21
12. 购物：单位价格　　　　　　　　　　　　　 23
13. 优惠券　　　　　　　　　　　　　　　　　 25
14. 网上购物　　　　　　　　　　　　　　　　 27
15. 小结　　　　　　　　　　　　　　　　　　 29
参考答案　　　　　　　　　　　　　　　　　　 32

Contents

INTRODUCTION
1. USING A SHOPPING LIST — 37
2. BUDGETING — 39
3. PARKING — 40
4. SALES TAX — 42
5. SPENDING CASH — 43
6. USING A DEBIT CARD — 45
7. USING A CREDIT CARD — 46
8. CLOTHES SIZING — 47
9. COST-PER-WEAR — 49
10. SALES AND DISCOUNTS — 50
11. BUY-ONE-GET-ONE SALES — 52
12. GROCERY SHOPPING: UNIT PRICE — 53
13. COUPONS — 54
14. ONLINE SHOPPING — 56
15. PUTTING IT ALL TOGETHER — 57
ANSWERS — 58

1
使用购物单

购物是米凯拉最喜欢的事情之一，无论买的是什么——衣服、生活用品或是图书。如果可以，她总是喜欢买最新的，也喜欢给她的亲朋好友选购礼物。

但有时她容易成为购物狂，在逛街时买得太多。为了防止这种疯狂购物的行为，米凯拉采取的措施是制订购物清单。这有助于她挑选真正需要的东西，避免过度消费。在购物清单上，她会估算出每一件物品要花费的金额，并在清单最后算出总额，这样就可以预估需要花多少钱。作为一个例子，下面列出了米凯拉去百货公司前准备的一份购物清单。

米凯拉的购物清单：

太阳镜，15美元
牛仔裤，20美元
笔记本，3美元
给爸爸的洗衣粉，4.50美元
给乔恩的生日礼物：电影光碟，20美元
给乔恩的生日卡片，2.50美元

米凯拉的预算以1美金或50美分为单位累加，这样更容易算出总和，以便更好地预估出这次购物的花费。

1. 根据估算，她一共要花费多少钱？

这时，米凯拉的爸爸提醒米凯拉要给他买新的耳机（有一次米凯拉不小心弄坏了爸爸的耳机，并保证会给他买新的），耳机预估25美元。

2. 现在米凯拉估计要花多少钱？

当米凯拉来到百货公司后，她发现有些物品的估价不太准确。下面是每件物品的实际价格：

太阳镜，16.53美元
牛仔裤，20.30美元
笔记本，2.57美元
洗衣粉，4.10美元
电影光碟，16.75美元
生日卡片，2.16美元
耳机，24.86美元

3. 米凯拉最终花了多少钱？

4. 哪一件物品是她估计得最接近于实际价格的？哪一件偏差最大？

2
预　算

在米凯拉结账前,她需要确认是否有足够的钱支付上述所有物品。因为她的资金有限,所以她事先做了预算。预算就是针对如何花钱做一个计划,包括有多少钱,需要买什么以及想要买什么。

米凯拉做了一份兼职工作,每周都能挣些钱。她把这笔收入的大部分存进了储蓄账户,而不是全花完。结余部分,她首先用来购买必需品,然后再买她喜欢的东西。

米凯拉估计她的钱不足以支付购物筐里的所有物品,请根据下面的内容,决定哪些物品她可以留着,哪些物品她需要放回购物架上。

米凯拉在脑海中列了如下清单：

必需品（必须买）：
笔记本，2.57美元
洗衣粉，4.10美元
电影光碟（礼物），16.75美元
生日卡片，2.16美元
耳机，24.86美元

喜欢的（非必须买）：
太阳镜，16.53美元
牛仔裤，20.30美元

1. 如果她只买必需品，她一共要花多少钱？

米凯拉每周给她的一个邻居做临时保姆，会挣40美元，外加帮另外一个邻居遛狗，每周通常4小时，每小时可以挣6美元。

2. 她每周能挣多少钱？

她会拿出40%的钱用于消费，其余的存在账户里供以后用。百分数指将总体分100份以后，这一部分占的份数。40%等于100份中的40份。要计算她每周收入的40%是多少，首先把40的小数点向左移两位，把它改成一个小数，然后按十进制乘法把它乘以米凯拉每周的总收入：

$$40\% = 0.40$$
$$0.40 \times 每周收入 = 每周可消费金额$$

3. 米凯拉每周会有多少钱可供消费？

现在，米凯拉有两周的可消费金额可以使用。

4. 现在她有多少钱？她是否有足够的钱支付购物清单上的必需品？

5. 她是否有足够的钱买一样她喜欢的东西？如果两样都需要买，她的钱够吗？

3
停 车

还有一次逛街，米凯拉和她的朋友们去购物中心。其中一个朋友珍妮有车，是她开车载她们一起过去的。她们绕着停车场来回找了很久，才找到了一个车位，并把车停了进去。但是当米凯拉拉开乘客那一侧车门时，发现车与车之间的间距太小了，自己没法出去。

即便在她们逛购物中心之前，仍然要用到数学知识。停车场是学习形状、大小和距离的一个非常好的地方。接下来的内容将帮你更好地理解停车场里的几何学。

当珍妮把车开进停车位时，她并没有意识到那里已经没有足够的空间让每位乘客都轻松地下车。

请帮珍妮找出足以让乘客和车都能出来的停车位面积。

矩形面积公式是：

$$A = 长度 \times 宽度$$

珍妮的车身长约为14英尺9英寸，宽约6英尺2英寸。在求面积时，你还应牢记1英尺等于12英寸。

1. 首先，把所有的长度和宽度都转化成以英寸为单位。

2. 然后，把数字代入面积公式。车的面积是多少平方英寸？

但是，这并不是她们需要的全部空间。她们还需要在车的两侧各留出1英尺的宽度以便乘客开门下车。这个额外空间应该加在车的宽度上，而长度保持不变。

3. 她们还需要多少英寸的额外空间？

4. 她们共需要多宽的停车位？

她们需要重新找一个更宽敞的停车位，你可以用同样的公式算出新车位的面积。

5. 她们需要的新车位面积是多少？这比汽车本身所占的面积大多少？

4
消费税

米凯拉和朋友们在第一个商店里就看到了中意的裙子。但是,这条裙子很贵,标价是60美元。米凯拉不确定自己是否有足够的钱,但是她很喜欢那条裙子。她宁可不买其他东西,也愿意把身上所有的钱用来购买这条裙子。

这时珍妮提醒她,付款时还需要另付消费税。消费税是政府在人们消费时收取的税费。政府通常用收取消费税和其他税费来为人们提供安全保障和公共服务,比如警察局、公园和学校。米凯拉所在州县的消费税率是7.3%。

现在米凯拉真不确定她是否买得起这条裙子了。米凯拉有足够的钱买这条裙子吗?我们可以通过下面的百分数找到答案。

米凯拉必须计算出她携带的所有钱，来看她自己是否可以买这条裙子。她把身上的钱拿出来数了数：

一张20美元纸币
两张10美元纸币
三张5美元纸币
8张1美元纸币
三枚25美分的硬币
两枚10美分的硬币

1. 米凯拉一共有多少钱？如果不算消费税的话，她的钱够买这条裙子吗？

现在我们需要算上裙子的消费税，消费税率是7.3%，我们有多种方法来计算总价。

首先，我们可以将税率写成小数形式，然后乘以裙子的标价，就像第2节的做法一样。现在，我们算出的仅仅是消费税，而不是总价。

2. 如何用小数表示7.3%？

3. 消费税是多少？

另一种理解百分数的方法是交叉相乘法。百分之一指的是100份中的一份，可以通过比较两个数找到它们之间的对应关系。

4.
$$\frac{7.3}{100} = \frac{X}{60}$$

$$100 \times X = 7.3 \times 60$$

$$X = (7.3 \times 60) \div 100$$

希望你使用这两种方法得到的结果是一样的！将消费税和裙子上的标价相加，我们就能得到裙子的总价。

5. 裙子的总价是多少？米凯拉买得起这条裙子吗？

5
现金消费

米凯拉购物的时候只用现金，她没有银行卡，因为她身上的现金就是她携带的所有钱，所以米凯拉决定不买这条裙子了。她没有足够的钱购买这条裙子，也不想向朋友们借钱。

这也意味着米凯拉还有60美元可以消费。她还想买牛仔裤和太阳镜。在进商场前，她列的购物清单上，这两件都被列为非必需品。现在她可以买哪件东西？她该如何分配她的钱？接下来，帮她解决这个问题吧。

米凯拉看到一条牛仔裤，标价为31.50美元。

1. 加上消费税后，裤子的总价是多少？

米凯拉买了这条牛仔裤，给了收银员一张20美元纸币，两张10美元纸币，三个25美分硬币和一个10美分硬币。

2. 收银员应该找给她多少零钱？

接下来，她想买太阳镜。可是，她又想到需剩一点儿钱和朋友们去买食品，她估计这需要7美元。

3. 米凯拉还剩下多少钱去买太阳镜呢？

她逛了几个商店，看到了一些标价为12.99美元的太阳镜（不含消费税）。

4. 太阳镜的消费税是多少？米凯拉有足够的钱买这副太阳镜吗？

米凯拉给了收银员2张10美元纸币，收银员找给她1张5美元纸币，3个25美分硬币和1个5美分硬币。

5. 收银员找给米凯拉的零钱够吗？如果不够，她应该再找多少？

购物结束了，米凯拉掏出了口袋里所剩的钱：

　　一张5美元纸币
　　三张1美元纸币
　　四枚25美分硬币
　　两枚10美分硬币
　　1枚1美分硬币

6. 米凯拉还剩多少钱？这些钱够她去买食品吗？

6
使用借记卡

米凯拉的朋友珍妮来商场时没有带现金,但是她带着一张借记卡。珍妮在银行开了一个支票账户,并把所有的钱存到了账户里。每当她用借记卡去买东西时,她就把钱存到支票账户里。

珍妮还可以使用借记卡从ATM机里取钱,或者查询账户余额。

通常情况下,想知道支票账户里还有多少钱可用来支付是很难的,这不像我们打开钱包就能知道还有多少现金。如果我们透支,银行还会收取一笔费用。这笔费用是我们必须缴纳的。使用借记卡和支票账户会使购物更加方便,但是也带来了一些责任。

珍妮也像米凯拉一样有个预算。她以为自己的支票账户里还有45美元。如果她的消费超过了45美元，银行将收取30美元的费用。

珍妮首先买了一对耳环，花了5.67美元（含消费税）。她非常肯定自己账户里的钱足够支付这对耳环。

她去的第二家商店只收现金，但是她想在那里买一条11.99美元的围巾。于是，珍妮只好去ATM机取现。她在ATM机那里查询到自己账户里只有23.12美元，这比她预想的要少。

1. 买耳环之前，珍妮账户里有多少钱？

珍妮至少想取出11.99美元，而ATM机里只能取20美元面值的钞票。

2. 珍妮能从ATM机里至少取多少钱？

这台ATM机是A银行设置的，但是珍妮的账户是B银行的。所以使用除A银行外的其他账户，还需要交3.50美元的手续费。

3. 珍妮将要从支票账户里取出多少钱（包括ATM机的手续费）？她账户里的钱够吗？

最终珍妮决定不取钱了。但是，在下一个可以使用借记卡的商店里，她找到了想要的东西（包含消费税），价格如下：

　　桌游卡，12.45美元
　　衬衫，25.98美元
　　钱包，10.20美元
　　电脑机箱，31.10美元

4. 哪些商品是珍妮可以买得起的？哪些是珍妮买不起的？

7
使用信用卡

与米凯拉一起购物的朋友小毅有一张信用卡。他有一份兼职工作,也喜欢阅读有关金融的资料。他非常喜欢使用信用卡,因为信用卡帮助他更好地理财和购物。

使用信用卡就如同借银行的钱。购物时,不必当天使用自己的钱来支付。信用卡提供的资金不是免费的。你必须在几周之后把花掉的钱还给银行。当你的钱不足以支付你想购买的商品时,信用卡就可以派上用场了。同时,你必须保证自己将来有能力偿还这笔钱,否则将会欠债。下面,我们一起来看看如何使用信用卡,再看看小毅是怎么做的。

小毅需要买一套西装参加校园舞会。尽管西装非常贵,但是他有一份兼职,而且他善于理财,所以小毅能够支付这笔钱。

问题是他现在没有足够的钱。他的支票账户里只有15美元。

但好消息是,他下周会收到80美元的薪水,下下周还有95美元的薪水。同时,他想从那两份薪水中各节省25美元,以备将来需要。

下面是小毅找到的一些西服，以及相应的标价：

西装1，99.99美元
西装2，220.00美元
西装3，115.30美元
西装4，170.00美元
西装5，105.60美元
西装6，199.99美元

1. 小毅现有的钱够买上面的一套西装吗？需要用信用卡吗？

2. 如果小毅使用接下来两周的薪水并且不用考虑节省钱，他能买得起上面的哪套西装？

3. 如果小毅只使用下周的薪水，能买得起哪套西装？

4. 如果小毅从接下来两周的薪水中各节省25美元，然后使用剩余钱来买西装，他买得起哪套西装？

5. 如果小毅将接下来两周薪水各节省25美元，并使用余款买第5套西装，那么他的薪水还能剩多少？不要忘了计算消费税。

8
衣服的尺码

米凯拉好久没有去逛商场了。她一直在攒钱准备买一条新的牛仔裤。虽然她还没想好买什么样的款式，但必须得买新的了，因为原来那条牛仔裤的膝盖那里快要磨破了。

她来到了商店，在琳琅满目的货架上，找到了自己喜欢的几个款式。每件衣服上都有尺码标识。可是，她看不懂那些标识上尺码的意思，不知道该去试穿多大号，于是她只好求助于售货员。

在买衣服的时候，我们常常需要在不同尺码标识体系中来回转换。美码体系中包含了女童、男童、青少年、男性、女性等各个系列。而采用欧码的服装，是基于人体的某部位尺寸来确定的。比如牛仔裤，欧码是基于腰围来确定的。

米凯拉通常穿的是美码青少年系列的衣服，所以她分不清那些美码女性系列和欧码标识的牛仔裤的实际尺寸。让我们来帮助她挑选适合她的衣服吧！

米凯拉通常穿美码青少年系列中的9号，偶尔也穿11号。而这家商店里的衣服标识只有美码的女性系列和欧码。

售货员告诉米凯拉，她可以将美码中的青少年系列转换为女性系列。青少年系列的尺码使用奇数，女性系列的尺码使用的是偶数。她只需要将自己平时穿的青少年系列尺码减去3，就可以得到相应女性系列的尺码。

1. 米凯拉应该穿女性系列中哪个尺码的衣服？

然后，售货员给米凯拉看了如下的尺码对照表，上面解释了美码青少年系列和欧码之间的换算：

美国青少年系列尺码	欧洲尺码
1	28
3	30
5	32
7	34
9	36
11	38
13	40
15	42

售货员还解释说欧码除了偶数尺码外，还有奇数尺码。例如，欧码33对应着美码青少年系列中的5 1/2。

售货员用软尺量出米凯拉的腰围是37.25英寸。

由于米凯拉的腰围不是一个整数，她需要四舍五入，找出最接近的尺码。

2. 根据尺码表，运用你的数学知识算出米凯拉可以试穿哪两个欧码尺寸？

9
衣服的性价比

现在，米凯拉已经找到了合适尺寸的牛仔裤，她准备试穿一下，然后挑选一条买下来。米凯拉选购牛仔裤首先考虑的是牛仔裤的性价比。米凯拉只能按商店的标价去支付。她每穿一次，就等于获得了一次使用价值。如果牛仔裤合身，又真的很喜欢，她就会穿很多次。即便是花了很多钱，只要能穿很多次，就是值得的。如果牛仔裤不合身，还不喜欢，她就会只穿几次，这样她就花了很多钱买了一件不怎么穿的衣服。

衣服的性价比是购买衣服时考虑的因素之一。你也可以将性价比用在其他方面。比如，一条毯子你使用多少次？一个视频游戏你玩多少次？你使用的次数越多，获得的使用价值就越多。知道如何计算衣服的性价比可以帮助你节省很多钱。

衣服性价比的计算方法是：

服装价格 ÷ 穿戴衣服的次数

衣服性价比的单位是美元/次。这真的很简单。

1. 如果你购买一件衬衫花了50美元，共穿了25次，那么性价比是多少？

米凯拉在店里逛了逛，找到三条她喜欢的牛仔裤。以下是这3条裤子含消费税的价格：

第1条，25美元
第2条，14.50美元
第3条，65美元

2. 这些牛仔裤的价格从低到高的顺序是什么？

米凯拉试穿了第1条牛仔裤，很合身，长度和款式也不错。她征询朋友路易莎的意见，路易莎也认为很合身。米凯拉觉得这条牛仔裤她每周穿一次，可以穿两年。

现在，请计算出第1条牛仔裤的性价比。首先，算出她将穿多少次，设定一年有52个星期，这会帮助你算出答案：

每周1次 × 52周 × 2年 = 104次

接下来，将上面的答案代入性价比公式。

3. 25美元 ÷ 104次 = ?

米凯拉又试穿了另外两条牛仔裤。第2条虽然便宜，但是不太合身，穿着不舒服。如果买这条，她可能只会每月穿一次，穿一年。第3条牛仔裤最贵，是一个名牌，和第1条一样合身。她觉得穿这条牛仔裤的次数会和第1条一样。

4. 计算出另外两条牛仔裤的性价比是多少？

米凯拉应该买性价比最高的牛仔裤。这意味着牛仔裤穿着成本最小，即性价比数值最低。

5. 基于性价比最高的原则，米凯拉应该买哪一条牛仔裤？为什么？

10
销售和折扣

回家之前,米凯拉还有一点时间,于是她去了书店。她喜欢阅读,但很久没有买书了。书店正在搞促销活动,几乎所有书都在打折!米凯拉还有富余的钱,她决定再买几本书。

促销活动中很多时候会涉及百分比和分数。一旦你弄清了百分比和分数,就能发现最大的优惠。接下来,让我们来了解一下它们在促销和折扣活动中的应用。

促销的图书上面贴了彩色标签，并用来区分它们的折扣是多少。以下是具体折扣情况：

红色：　9.5折
橙色：　9折
黄色：　7.5折
绿色：　5折
蓝色：　2.5折
紫色：　1折

米凯拉不用等到结账就能算出书的费用。把折扣比例换算成分数，她可以计算出它们的价钱。

一些常见的百分比和它们的对应分数是：

$$1/4 = 25\%$$
$$1/3 = 33\%$$
$$1/2 = 50\%$$
$$3/4 = 75\%$$

有时50%更容易理解，仅仅是原价的一半。

1. 请问贴着绿色标签标价12美元的书打折后的价格是多少？

2. 一本标价20美元贴着黄色标签的书折后价是多少？

有些百分比用口算是有点难度的，比如90%。但是，如果换个角度来看这个问题，你可以估算一个近似的促销价格。例如：

3. 如果一本书标价16美元，优惠90%，它的最终价格更接近3美元还是15美元？为什么？

4. 标价16美元的书，优惠90%后的实际价格是多少？用任意一种数学方法算出你的答案。

11
促销：买一送一

凯拉发现书店的某个区域正在进行买一送一的促销活动，在有的书架上，买一本书，就可以免费获得同书架上的另一本书；还有一些书架上，买一本书，可以半价获得同书架上的另外一本。

买一送一很划算，但有时候也会浪费你的钱。你原本只想买一样东西，却可能会被诱惑去买另一样东西，仅仅因为它在促销。

理解买一送一促销活动能帮你省钱，还可以让你买到需要的促销物品。下面我们来看看它的促销原理。

在买一送一的书架上，米凯拉发现一本她想要的小说，标价23美元，上面有一个蓝色的打折标签。还发现了一本想要的历史书，标价14美元，贴着橙色打折标签(请翻到20页查看不同颜色标签代表的折扣)。

1. 第一本小说要花多少钱？

2. 那本历史书要花多少钱？

在买一送一促销活动里，通常情况下，会免费赠送一个比较便宜的物品，但贵的那个物品是需要你付钱的。

3. 以上两本书中，哪一本是免费的？她一共要付多少钱？

4. 在打折和买一送一的促销活动中，她总共节省了多少钱？

在"第二件半价"的促销书架上，米凯拉只能找到一本她想要的小说，标价14.50美元，贴着绿色折扣标签。尽管如此，她还是想再买一本，这样就能享受到更多优惠。她又看到一本有些想要的手工书，标价19.25美元，贴着黄色折扣标签。

5. 这部小说要花多少钱？

6. 两本书中较便宜的那本只需半价，那么米凯拉买这两本书共花多少钱？

7. 你认为她应该买这两本书吗？为什么呢？

12
购物：单位价格

第二天米凯拉和爸爸去杂货店购物。在杂货店里，爸爸教她什么是单位价格。使用单位价格可以计算出买哪些食物最划算。

通常情况下，贴在货架上的标签，左上方会用明显的橙色标出食物的单位价格，即每单位食物要花多少钱。不同的食物用不同的度量单位，如磅、盎司、升、加仑等。

就价格而言，单位价格越低越好，这样的物品最值得购买。两个品牌相同的食品，你想知道哪一个更便宜，这时单位价格特别有用。在杂货店买东西要想省钱，一定要学会计算单位价格！

单位价格的计算公式是：

价格 ÷ 数量(单位)

例如，在麦片货架上，米凯拉和爸爸看到一盒24盎司的麦片，价格为3.69美元，一盒36盎司的麦片，价格为4.89美元。

买哪种更划算呢？哪种麦片更物美价廉呢？他们需要计算一下每种麦片的单位价格。以24盎司一盒的麦片为例，请你算算它的单位价格：

1. 3.69美元 ÷ 24盎司 = ？

2. 另一种麦片的单位价格是多少？

现在比较两种麦片的单位价格，较低的更值得购买，这表明购买同样多的麦片，花费更少。

3. 购买哪种麦片更划算？

如同上面的例子，与标价低的商品相比，那些标价高的商品有时实际上单位价格更低，购买更划算。

通常散装食品单位价格更便宜。当你需要买很多食物的时候，散装比小包装的食物的单位价格更便宜。当你学会了精打细算，就可以节省更多的钱了。

再来看一个例子。米凯拉想买一些杏仁。她直奔坚果货架，杏仁每包5.99美元，包装背面标的重量是8盎司。

然后她又去看了散装的杏仁。这里的杏仁是每磅7.99美元。如果她需要8盎司杏仁，应该选哪种？

你需要计算每盎司的单位价格。1磅等于16盎司，所以你要比较价格5.99美元8盎司的杏仁和价格7.99美元16盎司的杏仁哪种单位价格更低。

4. 哪一种的单位价格更低？应该买哪种？

13
优惠券

米凯拉的爸爸带来了一些杂货店的优惠券,有的是从报纸上剪下来的,有的是电子优惠券。优惠券就是可以随身携带的折扣券。商店的物品可能并没有打折,但是如果你有优惠券,就仍然可以得到折扣。

米凯拉的爸爸有几张优惠券。他想对自己要买的食物作比较,看看能省多少钱。接下来你可以算一算。

下面是他手里的杂货店优惠券：

麦片，优惠30%
果汁，买一送一
面包，买两个优惠20%
青豆罐头，优惠10%
冷冻浆果，第二件优惠40%

以下是他实际购买的食物及原价：

2盒麦片，每盒4.89美元
2瓶果汁，每瓶2.70美元
8盎司杏仁，4.05美元
2个面包，每个2.20美元
1听青豆，0.99美元
2袋冷冻浆果，每袋4.50美元
6个橙子，每个0.78美元

填写下面这张表来记录他购买的所有食物，及使用优惠券的情况：

物品	数量	初始单价/美元	优惠券	使用优惠券后价格/美元
麦片	2盒	4.89	7折	6.85

1. 米凯拉的爸爸总共要支付多少钱？

14
网上购物

凯拉最近在商店购买了很多东西。她还需要一些学习用品,但她不想再去实体店了,就从网上订购了一些。

网上购物可以让生活更便捷。你要做的就是在网上搜索到你所需要的东西,购买后等着送货上门就行了。很多在实体店购物的规则也同样适用于网上购物。送货上门服务需要你再支付一笔小额费用,这笔费用被称为"运费"。事实上,如果你花更多的精力去网上搜索的话,会买到比实体店更便宜的物品。

米凯拉有23美元现金，但还不够。她自己没有借记卡，所以不得不用妈妈的借记卡在网上购买一些学习用品。她又不想欠妈妈太多钱，所以她想将消费控制在预算内。

她需要三个笔记本，三个文件夹，几支笔，一个胶棒和一本学习计划手册。下面是她在网上搜到的学习用品：

笔记本，3美元
文件夹，0.75美元
一盒笔，4.80美元
胶棒，1.95美元
学习计划手册，6.99美元

1. 所有这些学习用品要花多少钱？

2. 这些学习用品加上7.3%的消费税后，要多少钱？

米凯拉将所有学习用品加入网上购物车，准备结账时，她才意识到自己忘了算运费！运费梯价表如下：

总订单0—10美元：运费为3.99美元
总订单10.01—20美元：运费为4.99美元
总订单20.01—50.00美元：运费为5.99美元

3. 她总共需要支付多少运费？

4. 她的订单总额是多少？她需要向妈妈借钱吗？

15
小　结

过去的几周，米凯拉进行过很多次的购物。期间她用了很多数学知识去计算预算、衣服尺码、折后价、单位价格等。下面我们来看看你是否还记得米凯拉购物时使用的这些数学知识，如果你学会了，当你去购物时也可以使用。

1. 如果每周你得到40美元，还想省下75%的钱，那么还剩多少钱可以花费？

2. 一个停车场是200英尺长、180英尺宽，它的面积是多少平方英尺？

3. 你想买的一款35.60美元的电视游戏。如果你所在州的消费税税率是8%，那么要交多少消费税？加上消费税后，这个电视游戏要花多少钱？

4. 你的银行支票账户还剩13美元，你还必须保证账户里至少有5美元。那么，你能买得起一个1.54美元的巧克力棒和一个10.50美元的游戏吗？

5. 如果她的腰围是38英寸，那么适合她的牛仔裤的欧洲尺码是多少？

青少年系列的衣服尺码是多少？

6. 如果你花34.80美元买了一件毛衣，共穿了73次，那么这件毛衣的性价比是多少？

7. 你在第二件半价的DVD碟片中挑选，找到两张想要的，一张标价13.50美元，另一张17美元。购买这两张DVD碟片需要花多少钱？

8. 你想买格兰诺拉燕麦卷时，发现其中一种是一盒6条，3.99美元，另一种是一盒10条，5.20美元。

考虑单位价格后，你应该买哪一种？

9. 你有一张能优惠25%的折扣券，购买一款售价为12.80美元的电视游戏需要花多少钱？

10. 妈妈让你用她的信用卡在网上订购一件售价为33.99美元的连衣裙，她说花费不能超过40美元，你认为这些钱足够了。但是你确定吗？消费税税率是7%，运费是3.50美元，这样的话你的钱还够吗？

参考答案

1.

1. 65美元
2. 90美元
3. 87.27美元
4. 耳机是估价最接近实际价格的物品(估高了0.14美元)，电影光碟的估计与实际价格差距最大(估高了3.25美元)

2.

1. 50.44美元
2. 64美元
3. 25.60美元
4. 是的，她的钱足够买清单上的必需品(25.60美元 × 2 = 51.20美元)
5. 不，她剩余的钱不够买她喜欢的任一件物品

3.

1. 长度=(14 × 12)+9 = 177英寸，宽度=(6 × 12)+2 = 74英寸
2. $A = 177 × 74 = 13098$平方英寸
3. 24英寸
4. 24+74 = 98英寸
5. $A = 177 × 98 = 17346$平方英寸；新车位面积比汽车本身所占面积大4248平方英寸

4.

1. 她共有63.95美元。如果不算消费税，她的钱足够买这条裙子
2. 0.073
3. 4.38美元
4. $X = 4.38$美元
5. 64.38美元；算上消费税，米凯拉的钱不够买这条裙子

5.

1. 31.50美元+(31.50美元 × 0.073) = 33.80美元

2. 7.05美元
3. 63.95美元-33.80美元-7美元= 23.15美元
4. 0.95美元；她的钱够买太阳镜(算上消费税，太阳镜总价为13.94美元)
5. 不够，还需要找给她0.26美元
6. 她还剩下9.21美元，所以足够她买食品

6.

1. 28.79美元
2. 20美元
3. 23.50美元；她账户里的钱不够
4. 买得起的商品：桌游卡，钱包；买不起的商品：衬衫，电脑机箱

7.

1. 不，他现有的钱不够，需要使用信用卡
2. 西装1，西装3，西装4，西装5
3. 买不起任一套
4. 西装1，西装3，和西装5
5. 105.60美元 +(105.60美元 × 0.073)= 113.31美元
   (80美元+95美元)-(25美元+25美元)-113.31美元= 11.69美元

8.

1. 6和08
2. 36和37

9.

1. 性价比是2美元/次
2. 第2条，第1条，第3条
3. 性价比是0.24美元/次
4. 第2条的性价比是1.21美元/次；第3条的性价比是0.63美元/次
5. 第1条，因为这条牛仔裤的平均每次穿着费用最低

10.

1. 6美元
2. 15美元

3. 3美元，因为90%是一个很大的折扣，意味着降价幅度非常大
4. 16美元 - (16美元 × 0.9) = 1.60美元

11.

1. 23美元 × 0.25 = 5.75美元
2. 14美元 × 0.9 = 12.60美元
3. 她将免费得到第一部小说，这两本书她一共要付12.60美元
4. 她将节省24.40美元
5. 14.50美元 ÷ 2 = 7.25美元
6. 19.25美 × 0.75 + 7.25美元 ÷ 2 = 18.06美元
7. 不应该买。她仅仅花7.25美元就可以买到自己真正想要的小说，但两本都买需要花18.06美元

12.

1. 0.15美元/盎司
2. 4.89美元 ÷ 16 = 0.14美元/盎司
3. 第二盒，即36盎司的麦片更划算
4. 包装版单价：5.99美元 ÷ 8盎司 = 0.75美元/盎司；散装版单价：7.99美元 ÷ 16盎司 = 0.50美元/盎司，她应该买散装杏仁

13.

1. 29.89美元

物品	数量	初始单价/美元	优惠券	使用优惠券后价格/美元
麦片	2盒	4.89	7折	6.85
果汁	2瓶	2.70	买一送一	2.70
杏仁	8盎司	4.05	无	4.05
面包	2条	2.20	8折	3.52
青豆	1听	0.99	9折	0.89
冷冻浆果	2袋	4.50	第二袋6折	7.20
橙子	6	0.78	无	4.68

14.

1. 24.99美元
2. 24.99美元 +(24.99美元 × 0.073) = 26.81美元
3. 5.99美元
4. 32.80美元；是的，她将欠妈妈一些钱

15.

1. 40美元 - (40美元 × 0.75) = 10美元
2. 200 × 180 = 36000平方英尺
3. 2.85美元；38.45美元
4. 买不起，因为支出将超过8美元
5. 38码；11码
6. 0.48美元
7. 23.75美元
8. 买第二种。第一种单价：3.99美元÷6条=0.67美元/条；第二种单价：5.20美元 ÷ 10条 = 0.52美元/条
9. 9.60美元
10. 是的，钱足够了。衣服、运费和消费税的总价是39.87美元

INTRODUCTION

How would you define math? It's not as easy as you might think. We know math has to do with numbers. We often think of it as a part, if not the basis, for the sciences, especially natural science, engineering, and medicine. When we think of math, most of us imagine equations and blackboards, formulas and textbooks.

But math is actually far bigger than that. Think about examples like Polykleitos, the fifth-century Greek sculptor, who used math to sculpt the "perfect" male nude. Or remember Leonardo da Vinci? He used geometry—what he called "golden rectangles," rectangles whose dimensions were visually pleasing—to create his famous *Mona Lisa*.

Math and art? Yes, exactly! Mathematics is essential to disciplines as diverse as medicine and the fine arts. Counting, calculation, measurement, and the study of shapes and the motions of physical objects: all these are woven into music and games, science and architecture. In fact, math developed out of everyday necessity, as a way to talk about the world around us. Math gives us a way to perceive the real world—and then allows us to manipulate the world in practical ways.

For example, as soon as two people come together to build something, they need a language to talk about the materials they'll be working with and the object that they would like to build. Imagine trying to build something—anything—without a ruler, without any way of telling someone else a measurement, or even without being able to communicate what the thing will look like when it's done!

The truth is: We use math every day, even when we don't realize that we are. We use it when we go shopping, when we play sports, when we look at the clock, when we travel, when we run a business, and even when we cook. Whether we realize it or not, we use it in countless other ordinary activities as well. Math is pretty much a 24/7 activity!

And yet lots of us think we hate math. We imagine math as the practice of dusty, old college professors writing out calculations endlessly. We have this idea in our heads that math has nothing to do with real life, and we tell ourselves that it's something we don't need to worry about outside of math class, out there in the real world.

But here's the reality: Math helps us do better in many areas of life. Adults who don't understand basic math applications run into lots of problems. The Federal Reserve, for example, found that people who went bankrupt had an average of one and a half times more debt than their income—in other words, if they were making $24,000 per year, they had an average debt of $36,000. There's a basic subtraction problem there that should have told them they were in trouble long before they had to file for bankruptcy!

As an adult, your career—whatever it is—will depend in part on your ability to calculate mathematically. Without math skills, you won't be able to become a scientist or a nurse, an engineer or a computer specialist. You won't be able to get a business degree—or work as a waitress, a construction worker, or at a checkout counter.

Every kind of sport requires math too. From scoring to strategy, you need to understand math—so whether you want to watch a football game on television or become a first-class athlete yourself, math skills will improve your experience.

And then there's the world of computers. All businesses today—from farmers to factories, from restaurants to hair salons—have at least one computer. Gigabytes, data, spreadsheets, and programming all require math comprehension. Sure, there are a lot of automated math functions you can use on your computer, but you need to be able to understand how to use them, and you need to be able to understand the results.

This kind of math is a skill we realize we need only when we are in a situation where we are required to do a quick calculation. Then we sometimes end up scratching our heads, not quite sure how to apply the math we learned in school to the real-life scenario. The books in this series will give you practice applying math to real-life situations, so that you can be ahead of the game. They'll get you started—but to learn more, you'll have to pay attention in math class and do your homework. There's no way around that.

But for the rest of your life—pretty much 24/7—you'll be glad you did!

1
USING A SHOPPING LIST

One of Mikayla's favorite things to do is to go shopping. It doesn't even matter what she's buying—clothes, groceries, books. Mikayla likes buying new stuff when she can, and she really likes picking out and buying presents for her family and friends.

Sometimes she can end up going overboard, though, and buying too much during a shopping trip. One of the things Mikayla does to help her not do that is make a shopping list. Making a shopping list helps keep her focused on getting what she needs, and not spending too much money. On her list, she estimates how much each item will cost, and then finds the total at the bottom. Then she knows how much she expects to spend. The next page shows an example of a shopping list Mikayla made before going to a department store.

Mikayla's shopping list:

Sunglasses, $15
Jeans, $20
Notebook, $3
Laundry detergent for Dad, $4.50
Present for Jon: Movie, $20
Birthday card for Jon, $2.50

All of Mikayla's estimates are either whole dollars, or dollars with 50 cents added on. That makes it easy to add up the estimates to get a good idea of how much she will spend on her shopping trip.

1. What is the total she estimates she will spend?

Then her dad reminds her to buy him some new headphones. She accidentally broke his, and promised to buy him some new ones. The headphones will cost $25.

2. Now how much does Mikayla estimate she will spend?

When Mikayla goes to the store, she finds that not all her estimates were exactly right.

This is what each of the items actually costs:

Sunglasses, $16.53
Jeans, $20.30
Notebook, $2.57
Laundry detergent, $4.10
Movie, $16.75
Birthday card, $2.16
Headphones, $24.86

3. How much will she end up paying?

4. Which of her estimates was closest? Which was farthest off?

2
BUDGETING

Before Mikayla checks out, she has to see if she can afford to buy everything. She is on a budget, because she doesn't have unlimited supplies of money! A budget is a plan of how to spend money. It includes how much money you have, what you need to buy, and what you want to buy.

Mikayla has a part-time job and makes a little money each week. She puts a lot of it in her **savings account**, though, so she doesn't spend it all. The rest she uses to first buy what she needs, and then what she wants.

Mikayla suspects she can't afford to buy everything in her basket. Look at the next page to decide what she can keep and what she has to put back.

Mikayla makes a list in her head that looks like this:

Needs (Have to buy)
Notebook, $2.57
Laundry detergent, $4.10
Movie (present), $16.75
Birthday card, $2.16
Headphones, $24.86

Wants (Don't have to buy)
Sunglasses, $16.53
Jeans, $20.30

1. If she only buys what she needs, how much would she spend?

Mikayla makes $40 babysitting for a neighbor every week, plus $6 an hour for walking another neighbor's dog, which she usually does for 4 hours a week.

2. How much does she make every week?

She sets aside 40% of her money to spend and puts the rest in her savings account for later.

Percents are parts out of 100. 40% is like saying 40 parts out of 100. To figure out how much 40% of her weekly income is, first move the decimal point in 40 over to the left two places to change it into a decimal number. Then multiply how much Mikayla makes a week by the decimal:

40% = .40
.40 x weekly income = spending money

3. How much does she have to spend every week?

Right now, Mikayla has two weeks' worth of money to spend.

4. How much does she have? Does she have enough to buy what she needs on her shopping list?

5. Does she have enough to also buy one of the items she wants? Could she buy both of them?

3
PARKING

On another shopping trip, Mikayla and her friends go to the mall. One of her friends, Janine, has a car and drives them all to the mall. They circle around the parking lot for a while, looking for a space. They finally find one and pull in. But when Mikayla goes to open the passenger side door, she can't get out! The space is too narrow.

Even before they step foot into the mall, math is involved in their shopping trip. The parking lot is a great place to explore geometry, the math of size and space. The next pages will give you a better understanding of parking-lot geometry.

When Janine pulled into the parking space, she didn't realize there wasn't quite enough room for everyone to get out easily.

Find the area of the space Janine needed to park in so that there was enough room for both the car and the passengers getting out.

The equation for the area of a rectangle is:

$$A = \text{length} \times \text{width}.$$

Janine's car is approximately 14 feet and 9 inches long and 6 feet and 2 inches wide. To find the area, you'll need to remember there are 12 inches in 1 foot.

1. First, convert the length and width to inches:

2. Next, plug those numbers into the area formula. What is the area of the car in square inches?

But that isn't all the room they need. They also need 1 foot of space on either side of the car to be able to open the doors and get out. This extra space has to be added to the width of the car. The length will stay the same.

3. How much extra space in inches do they need?

4. What is the total width they now need for the parking space?

You can find the new, bigger area they will need for the parking space using the same area equation.

5. What is the new area they need? How much bigger is it than the area the car takes up?

4
SALES TAX

At the first store Mikayla and her friends go into, Mikayla finds the perfect dress. It's pretty expensive, though. The tag says it is $60. Mikayla isn't sure she has enough money, but she really loves the dress. She is willing to spend all the money she brought on it, even though it would mean she couldn't buy anything else.

Then Janine reminds her that she has to add in sales tax to the price on the tag. Sales tax is the money the government collects on purchases. The government uses the sales tax money—and the other taxes it collects—to provide protection and services to people, like police, parks, and schools. The sales tax rate in Mikayla's state and county is 7.3%.

Now Mikayla really isn't sure if she can buy the dress. Should Mikayla buy the dress? Does she have enough money? You can figure it out with percents on the next page.

Mikayla has to count all the money she brought with her to know if buying the dress is even an option. She takes out her cash and counts:

One $20 bill
Two $10 bill
Three $5 bills
Eight $1 bills
3 quarters
2 dimes

1. How much does he have in total? Does she have enough to buy the dress without the sales tax?

Now figure out how much the dress would be with the sales tax added on. You know the tax is 7.3%. You can use several ways to calculate the price.

First, you can move the decimal point and multiply like you did in Chapter 2. Right now, you are just finding the amount of sales tax, not the total price of the dress.

42

2. What is 7.3% in decimal form?

3. What is the sales tax?

Another way to use percents is to cross multiply. A percent is a part out of 100, which you can compare to the relationship between the numbers you are using.

4.
$$\frac{7.3}{100} = \frac{X}{60}$$

$$100 \times X = 7.3 \times 60$$

$$X = (7.3 \times 60) \div 100$$

Hopefully, you got the same answer using both methods! Now you can add the sales tax to the price tag and see what the total price for the dress is.

5. What is the total price? Can Mikayla afford it?

5
SPENDING CASH

Mikayla only uses cash to buy things when she goes shopping. She doesn't have any bank cards. Because her only money is the cash she has with her, she decides not to buy the dress. She can't afford it, and she doesn't really want to borrow money from her friends.

That means she still has $60 to spend if she wants. She still wants to buy jeans and sunglasses, which are the items she couldn't afford to buy on her shopping list when she went to the department store earlier. What can she buy? How should she divide up her money? Help her figure it out on the next page.

Mikayla finds a pair of jeans that have a tag saying $31.50.

1. How much are the jeans, including sales tax?

She buys the jeans and gives the cashier a $20 bill, two $10 bills, 3 quarters, and a dime.

2. How much change should she get?

Now she just wants to find sunglasses. However, she also wants to save a little bit to buy something in the food court with her friends. She estimates she will need $7 left over.

3. How much money does she have left to buy the sunglasses?

She shops around at a few more stores and finds some sunglasses for $12.99, not including tax.

4. How much will the sales tax be? Does she have enough money to buy the sunglasses?

Mikayla gives the cashier two $10 bills. The cashier hands back a $5 bill, three quarters, and a nickel.

5. Did she give back enough change? If not, how much more does she need to give Mikayla?

After her purchases, Mikayla reaches into her pocket and takes out all the money she has left. She has:

one $5 bill
three $1 bills
four quarters
two dimes
a penny

6. How much does she have left? Is it enough to buy something at the food court?

6
USING A DEBIT CARD

Mikayla's friend Janine doesn't bring cash to the mall. Instead, she has a debit card. She set up a checking account with the bank, where she keeps all her spending money. Every time she **deposits** money in her checking account, she can use her debit card to make purchases.

She can also use her debit card with an ATM (Automated Teller Machine). She can **withdraw** cash from her account if she ends up needing it, or she can check how much money she has left in her account.

It's often harder to tell how much money you have left to spend in a checking account. You can't just look in your wallet like you can with cash. If you spend too much money, the bank will charge you a **fee**, and then you'll have to pay the bank. Having a debit card and checking account can make shopping easier, but they also come with some responsibilities!

Janine, like Mikayla, is also on a budget. She thinks she has $45 left in her checking account. If she spends more than that, the bank will charge her $30 for using more money than is in her account.

The first thing Janine buys is a pair of earrings for $5.67, with tax. She's pretty sure she has enough money for them in her account.

The next store she goes into takes cash only, and she wants to buy a scarf for $11.99. Janine only has her card, but she goes to an ATM for some cash. While she's there, she checks her balance, to see how much money she really has left. The ATM says she has $23.12 in her account, which is less than she thought she had.

1. What was her balance before she bought the earrings?

She wants to take out at least $11.99. The ATM only gives out money in **multiples** of $20.

2. What is the least amount of money Janine can take out of the ATM?

The ATM is also run by Bank A, but Janine has an account at Bank B. The ATM charges anyone not from Bank A $3.50 for using it.

3. How much money will she be taking from her checking account, including the ATM fee? Does she have enough money in her account?

In the end, she doesn't take out any money. But in the next store, which does take debit cards, she sees these things she wants (including sales tax):

Board game, $12.45
Shirt, $25.98
Wallet, $10.20
Computer case, $31.10

4. Which of these things can she afford, and which can she not afford?

7
USING A CREDIT CARD

Mikayla's friend Yi, who is also shopping with them, has a credit card. He has a part-time job, and likes to read about **finances**. He likes having a credit card, because it helps him understand money and shopping a little better.

Using a credit card is like borrowing money. You can make a purchase today and not have to use your own money to make your purchase. However, credit cards are not free money. You will have to pay the bank back in a few weeks. Credit cards are useful when you don't have enough money to buy something you really need right now, but you know you will have money for it a little later on. If you can't pay for your purchase later, though, you will go into debt. Find out how credit cards really work, and how Yi uses his, on the next page.

Yi needs to buy a new suit for a school dance. Suits are pretty expensive, but he likes to pay for things himself because he has a part-time job, and he likes to **manage** his money.

The trouble is, he doesn't have enough money right now. He only has $15 in his checking account.

The good news is that he is getting a paycheck next week for $80, and one the week after for $95. He likes to save $25 from each paycheck for the future.

Here are the suits Yi finds, and their price tags:

Suit 1: $99.99
Suit 2: $220.00
Suit 3: $115.30
Suit 4: $170.00
Suit 5: $105.60
Suit 6: $199.99

1. Can Yi pay for any of the suits with the money he has right now? Will he need to use his credit card?

2. Which of the suits could Yi pay off if he used all the money from both his paychecks and didn't save any money?

3. Which of the suits could Yi pay off if he used only the money from his paycheck next week?

4. Which of the suits could Yi pay off if he used only the money that was left from both paychecks after he put some aside for savings?

5. How much money would Yi have left from his paychecks if he put some aside for savings and also bought Suit 5? Don't forget to add on sales tax.

8
CLOTHES SIZING

Mikayla hasn't gone shopping in a few weeks. She has been saving up her money to buy some new jeans. She isn't quite sure what she wants yet, but her old jeans are wearing thin in the knees. It's time to buy some new ones.

When she gets to the store, she starts looking through the racks. She finds a couple styles she really likes and looks at the size tag. They have sizes on them she doesn't recognize! She isn't sure which ones she should take to the dressing room, so she goes and finds a store employee.

47

Sometimes when you are shopping for clothes, you will need to **convert** from one size system to another. In the United States, people may wear girls' sizes, boys' sizes, juniors' sizes, men's sizes, women's sizes, and more! Some clothes come in European sizes too, which are based on measurements. For jeans, European sizes are based on how big around your waist is in centimeters, which is the circumference of (distance around) your waist.

Mikayla usually wears juniors' sizes, but she has stumbled on jeans with women's and European sizes. Help her figure out what size she should try on.

Mikayla usually wears a size 9 in juniors' sizing, sometimes an 11. She can't find any juniors' sizes in this store, though, only women's and European sizes.

The store employee tells Mikayla she can convert from juniors' sizes to women's sizes in her head. Juniors' sizes come in odd numbers, and women's sizes come in even numbers. She just needs to subtract 3 from the juniors' size she normally wears.

1. Which women's sizes should Mikayla try on?

Then the store employee shows Mikayla the following sizing cart, to explain how European sizing works:

U.S. Juniors' size	European size
1	28
3	30
5	32
7	34
9	36
11	38
13	40
15	42

She explains that there are also odd numbers in between the even European sizes. So, a size 33 jean would **correspond** to a size 5½ in juniors' sizes.

The store employee takes out a tape measure and measures Mikayla's waist, which is 37.25 inches around.

Because Mikayla's waist isn't a whole number, she should round up and down to the nearest whole numbers to find out which sizes she should try on.

2. According to the chart, and your own math, which two European sizes should she try on?

9
COST-PER-WEAR

Now that Mikayla has found the right jean size, she's ready to try on some and buy a pair. One way Mikayla can pick a pair of jeans is to think about the cost-per-wear. Mikayla will have to pay however much the jeans cost right at the store. Every time she wears them after that, she is getting some value out of them. If the jeans fit and she really likes them, she will wear them a lot. She might have paid a lot of money, but it was worth it because she wears them a lot. If the jeans don't fit and she doesn't like them, she will wear them only a couple times. She just paid a lot of money for something she doesn't wear.

Cost-per-wear is a way of thinking about buying clothes. You could also think of other items in terms of cost-per-use. How many times do you use a blanket? Or a video game? The more you use them, the more value you're getting out of them. Knowing how to calculate cost-per-wear can help you save lots of money over time.
The equation for cost-per-wear is:

price of the clothing ÷ how many times you wear the clothing

The units of cost-per-wear are dollars per wear. It's really very simple.

1. What would the cost-per-wear be if you paid $50 for a shirt, and you wore it 25 times?

Mikayla shops around for some jeans and finds 3 pairs she likes. Here are their prices, including sales tax:

Pair 1: $25
Pair 2: $14.50
Pair 3: $65

2. What is the order of these jeans from cheapest to most expensive?

Mikayla tries on Pair 1. They fit really well, and are the right length and style. She asks her friend Luisa's opinion, and she thinks they fit well too. Mikayla thinks she would wear them once a week for two years before they wore out.

Now figure out the cost-per-wear for Pair 1. First, figure out how many times she would wear them before they wore out. There are 52 weeks in a year, which will help you figure it out.

$$1 \text{ time a week} \times 52 \text{ weeks} \times 2 \text{ years} = 104 \text{ times}$$

Now plug the numbers into the cost-per-wear equation.

3. $\$25 \div 104 \text{ times} =$

Mikayla tries on the other two pairs. Pair 2 is cheap, but uncomfortable and doesn't quite fit right. If she bought them, she might only wear them once a month for a year. Pair 3 is really expensive and a designer brand, and fits about as well as Pair 1. She thinks she would wear them about the same amount.

4. What are the cost-per-wear numbers for the other two pairs?

Mikayla should buy the jeans with the best cost-per-wear value. That means the jeans that cost the least whenever she wears them—the lowest cost-per-wear number.

5. Which pair of jeans should Mikayla buy based on cost-per-wear? Why?

10
SALES AND DISCOUNTS

Mikayla still has time to shop before she has to go home, so she heads over to the bookstore. She likes to read, but hasn't bought any new books in a while. The bookstore is having a big sale. Almost everything is discounted! Mikayla decides she has enough money to buy a few books.

Sales have a lot to do with percents and fractions. Once you figure out percents and fractions, you'll be able to find the best sales. Check out the next page to practice with sales and discounts.

The books on sale have colored labels on them that identify how much they're discounted. Here's the system:

Red: 5% off
Orange: 10% off
Yellow: 25% off
Green: 50% off
Blue: 75% off
Purple: 90% off

Mikayla doesn't necessarily want to wait till she gets to the cash register to find out how much the books cost. She can calculate their prices if she can change the percents into fractions.
Some common percents and their fraction **equivalents** are:

¼ = one-fourth = 25%
⅓ = one-third = 33%
½ = one-half = 50%
¾ = three-quarters = 75%

Sometimes it's easier to understand that 50%, for example, is just half of the price.

1. Can you figure out how much a book with a green tag that is $12 would cost after the discount?

2. How about a book with a yellow tag that costs $20?

Some of the percents are a little harder to calculate off the top of your head, like 90%. But if you ask yourself good questions, you can arrive at an approximate sales price. For example:

3. If book is $16 and 90% off, will it be more likely to cost $3 or $15? Why?

4. What will a $16 book that is 90% off really cost after the discount? Use any math you want to calculate the answer.

11 BUY-ONE-GET-ONE SALES

Mikayla finds a section of the bookstore that is selling books using a buy-one-get-one sale. On one shelf, if she buys one book, she gets another one on the same shelf free. On another shelf, if she buys one book, she gets another half off.

Buy-one-get-one sales can be good deals, but they can also make you waste your money. If you really only want to buy one thing, you might get tricked into buying another if you know you can buy it on sale.

Understanding buy-one-get-one sales can help you save money and get things you really need and want on sale. Turn to the next page to figure out how they work.

On the buy-one-get-one shelf, Mikayla finds a novel she really wants. It costs $23, but it has a blue sticker on it, so it's also on sale. She also finds a history book she wants to read, and it costs $14 with an orange sticker. (Turn back to page 55 to see what the discounts are for each color sticker.)

1. How much does the first novel cost?

2. How much does the history book cost?

In buy-one-get-one-free sales, you will usually get the item that is cheaper for free. You will have to pay for the one that is more expensive.

3. Which book will she get for free? How much will she pay for both books?

4. How much money does she save in total with the discounts and the buy-one-get-one-free deal?

Now Mikayla looks at the buy-one-get-one-half-off shelf. On that shelf, Mikayla can only find one book she really wants to read, a novel that costs $14.50 and has a green sticker on it. She feels like she should buy something else, though, so she can take advantage of the deal. She sees a craft book she sort of wants, which is $19.25 and has a yellow sticker.

5. How much does the novel cost?

6. How much would buying both books cost? Mikayla will get half off of the book that costs less.

7. Do you think she should buy both books? Why or why not?

12 GROCERY SHOPPING: UNIT PRICE

Mikayla goes grocery shopping with her dad the next day. While at the grocery store, he teaches Mikayla about something called unit price. He uses the unit price all the time to figure out which foods are better values, and which he should buy.

You can find the unit price of a food usually in the upper left part of the shelf label, often highlighted in orange. It tells you how much the food is per unit. The unit is a measurement that varies from food to food. The unit could be pounds, ounces, liters, gallons, or something else.

The best unit prices are the cheapest. They show what the best buy will be, in terms of price. Unit prices are especially helpful if you are trying to compare two brands of the same food and want to get the one that is truly cheaper. If you're trying to save money at the grocery store, unit prices are really helpful!

The equation for unit price is:

$$\text{cost} \div \text{quantity (in units)}$$

For example, Mikayla and her dad take a look at the cereal aisle. They see one 24-ounce box of cereal for $3.69 and one 36-ounce box for $4.89.

Which one is the better deal? Which one will give them more ounces of cereal for a lower price? They need to find the unit price for each one. The equation for the 24-ounce box looks like this. Find the answer:

1. $3.69 ÷ 24 ounces =

2. What is the unit price for the second box?

Now compare the two unit prices. The one that is lower is the better deal. Mikayla and her dad will pay less for the same amount of cereal.

3. Which one is the better deal?

Sometimes, like in that example, an item that looks more expensive in price will be a better deal. But you're paying less per ounce than the option that looks cheaper.

Often, food in **bulk** will have a better unit price. You have to buy a lot of food, but it will be cheaper per unit than food you buy in smaller packages. If you know how to shop smart, you can save money.

Try another example. Mikayla wants to buy some almonds. She heads over to the nut shelf and sees that a package of almonds is $5.99. She looks at the back of the package and sees there are 8 ounces of almonds in the package.

Then she goes to the bulk section to look at almonds. Those almonds are $7.99 a pound. Which almonds should she buy if she wants 8 ounces?

You will need to find the unit price in cost per ounces. There are 16 ounces in a pound, so you are comparing 8 ounces of almonds at $5.99 and 16 ounces of almonds at $7.99.

4. Which ones have a cheaper unit price? Which ones should she buy?

13
COUPONS

Mikayla's dad has brought along some coupons to the grocery store. He clipped them out of the newspaper, and he also found a couple online. Coupons are like discounts you carry around in your pocket. The items in the store might not be advertised as on sale, but if you have a coupon, you can still get a discount.

Mikayla's dad has several coupons. He wants to match them up to the food he's buying and see how much he is going to save. You can do the math too, on the next page.

Here are all the coupons he brought with him to the grocery store:

cereal, 30% off
juice, buy-one-get-one-free
bread, 20% off two loaves
canned green beans, 10% off
frozen berries, buy one get one 40% off

And here's what he actually bought:

2 boxes cereal, $4.89 each
2 bottles juice, $2.70 each
8 ounces almonds, $4.05
2 loaves bread, $2.20 each
1 can green beans, $.99
2 bags frozen berries, $4.50 each
6 oranges, $.78 each

Fill in this chart to keep track of everything he bought, and all the coupons:

Item	Amount	Original price (each)	Coupon	Price for all after coupon
Cereal	2 boxes	$4.89	30% off	$6.85

1. What was the total that Mikayla's dad paid at the cash register?

14
ONLINE SHOPPING

Mikayla has done a lot of shopping in stores lately. She needs some school supplies, but she doesn't really feel like going out to the store again. Instead, she orders what she needs online.

Online shopping can make life easier. All you have to do is search on your computer for exactly what you need, and then have it sent directly to your door. A lot of the same rules that apply to shopping in stores also apply to shopping online. And you also have to remember you'll pay a little extra for having it sent to you. This extra cost is called "shipping." However, you might be able to find cheaper items online than in the store if you search hard enough.

Mikayla has $23 in cash. She will have to use her mom's debit card to buy her school supplies online, because she doesn't have one of her own. She doesn't want to owe her mom too much money, though, so she is trying to stick to her budget.

She needs three notebooks, three folders, some pens, a binder, and a schedule planner. These are the options she finds online:

notebook, $3
folder, $.75
package of pens, $4.80
binder, $1.95
planner, $6.99

1. How much does everything cost?

2. How much does everything cost with the 7.3% sales tax added on?

Mikayla puts everything in her shopping cart online and goes to the checkout. As she's checking out, she realizes she forgot about shipping! She sees the shipping rates listed:

total order $0–$10: $3.99
total order $10.01–$20: $4.99
total order $20.01–$50.00: $5.99

3. How much will she have to pay in shipping?

4. How much is her total order? Will she owe her mom any money?

15
PUTTING IT ALL TOGETHER

Mikayla has done a lot of shopping over the last few weeks. She has also used a lot of math to figure out budgets, determine her size in clothes, calculate sale prices, use unit prices, and more. See if you can remember some of the shopping math Mikayla has used, and which you can use too when you go shopping.

1. If you make $40 a week and you want to save 75% of it each week, how much do you have left over to spend?

2. A parking lot is 200 feet long and 180 feet wide. What is its area in square feet?

3. A video game you want to buy is $35.60. What will the sales tax be on it if the tax in your state is 8%? How much will the video game cost with tax?

4. You have $13 left in your checking account in the bank, and you have to keep at least $5 in there at all times. Can you afford to buy a chocolate bar that costs $1.54 and a game that costs $10.50?

5. What European size jeans would someone wear if her waist were 38 inches in circumference?

What size juniors' clothes would she wear?

6. What is the cost-per-wear if you buy a sweater for $34.80 and wear it 73 times?

7. You are looking at a bunch of DVDs that are buy-one-get-one-half-off. You find two you want—one is $13.50 and the other is $17. How much will you spend on both of them?

8. You are trying to decide whether to buy a box of granola bars that are $3.99 for 6 bars, or another box that is $5.20 for 10 bars.

    Which should you buy, based on unit price (price per bar)?

9. You have a 25% off coupon for buying a video game that costs $12.80. How much will the game cost?

10. Your mother is letting you use her credit card to order a dress online that costs $33.99. She says you can't spend more than $40, and you figure you have plenty of money. But are you sure? Your sales tax is 7%, and the shipping is $3.50. Do you have enough money?

Answers

1.

1. $65
2. $90
3. $87.27
4. The headphones were the closest ($.14 off) and the movie was the farthest off ($3.25).

2.

1. $50.44
2. $64
3. $25.60
4. Yes ($25.60 x 2 = $51.20)
5. No, she can't afford to buy either of them.

3.

1. Length = (14 x 12) +9 = 177 inches, Width = (6 x 12) +2 = 74 inches.
2. A = 177 x 74 = 13,098 square inches
3. 24 inches
4. 24 + 74 = 98 inches
5. A = 177 x 98 = 17,346 square inches; the new area is 4,248 square inches larger.

4.

1. Yes, she has $63.95
2. .073
3. $4.38
4. X= $4.38
5. $64.38; no she can't quite afford it.

5.

1. ($31.50) + ($31.50 x .073) = $33.80
2. $7.05
3. $63.95 – $33.80 – $7 = $23.15
4. $.95; Yes (the total will be $13.94)
5. No, she needs to give her $.26 more.
6. She has $9.21 left, so yes, she has enough.

6.

1. $28.79
2. $20
3. $23.50; no, she doesn't have enough money.
4. Afford: board game, wallet; Not afford: shirt, computer case

7.

1. No, he can't pay for any so he will need to use his credit card.
2. Suit 1, Suit 3, Suit 4, and Suit 5
3. None.
4. Suit 1, Suit 3, and Suit 5.

5. $105.60 + ($105.60 x .073) = $113.31
   ($80 + $95) − ($25 + $25) − $113.31 = $11.69

8.

1. 6 and 8
2. 36 and 37

9.

1. $2 per wear
2. Pair 2, Pair 1, Pair 3
3. $.24 per wear
4. Pair 2: $1.21 per wear, Pair 3: $.63
5. Pair 1, because they are cheapest cost-per-wear.

10.

1. $6
2. $15
3. $3, because 90% off is a big discount and means a lot of money will be taken off the price.
4. $16 − ($16 x .9) = $1.60

11.

1. $23 − ($23 x .75) = $5.75
2. $14 − ($14 x .10) = $12.60
3. She will get the first novel for free; she will pay $12.60 for both books.
4. She will save $24.40
5. $14.50 ÷ 2 = $7.25
6. $19.25 − ($19.25 x .25) + ($7.25 ÷ 2) = $18.06
7. Probably not. She would only spend $7.25 if she bought the novel she really wanted, but she would spend $18.06 if she bought both books.

12.

1. $.15 per ounce
2. $4.89 ÷ 16 = $.14 per ounce

3. The second box with 36 ounces
4. Packaged: $5.99 ÷ 8 ounces = $.75; Bulk: $7.99 ÷ 16 ounces = $.50; she should buy the bulk almonds.

13.

1. $29.89

Item	Amount	Original price (each)	Coupon	Price for all after coupon
Cereal	2 boxes	$4.89	30% off	$6.85
Juice	2 bottles	$2.70	buy-one-get-one-free	$2.70
Almonds	8 ounces	$4.05	none	$4.05
Bread	2 loaves	$2.20	20% off	$3.52
Green beans	1 can	$.99	10% off	$.89
Frozen berries	2 bags	$4.50	buy-one-get-one 40% off	$7.20
Oranges	6	$.78	none	$4.68

14.

1. $24.99
2. $24.99 + ($24.99 x .073) = $26.81
3. $5.99
4. $32.80; yes, she will owe her mom money.

15.

1. $40 − ($40 x .75) = $10
2. 200 x 180 = 36,000 square feet
3. $2.85; $38.45
4. No, you will be spending more than $8.
5. Size 38; 11
6. $.48
7. $23.75
8. Buy the second box ($3.99 ÷ 6 = $.67 per bar or $5.20 ÷ 10 = $.52 per bar).
9. $9.60
10. Yes. The dress, shipping, and sales tax added together comes to $39.87.